Planetary Grid Systems

by

Tenzin Gyurme

ISBN No. 978-0-557-500-34-5

This book is dedicated to

The Seeker on the Path.

The Path to liberation from

The Human Conditioning.

And as part of our conditioning is

Listening to an artifice that

Was created by Man;

Then releasing the hold that Artifice has,

Is the key to stepping into a new Reality.

A Galactic Reality.

CONTENTS

Introduction

Has modern technology and its 'conveniences' really made us an aware and advanced race?

Or has the artifices that we have overlaid on a galactic reality just enslaved us to mediocrity?

Does our time keeping system really work?

Or are we slaves to an artificial constant created by an Emperor in Rome?

(see www.lawoftime.org for a handbook that will release you from that fraud)

Does our system of Laws in place on this planet really work?

Or are we slaves to an artificial constant created by an Emperor in Rome?

(see '*All of the Universal Laws and Principles Governing Life*' available at www.houseofgoddess.eu for more information to release you from that fraud)

Or worse yet, are you believing the blasphemy published by the Emperor of the Holy Roman Empire?

If you are, you might want to read, *Our Christ-ening, The Book of Enoch and The Gnostic Books of Formation* (also available on the above website) for the '*real*' story.

How about our system of Math?

Are you still believing either the Roman numbering or Arabic numbering system, currently in place?

If so, could you once again, be falling for the tricks of the Emperor in Rome?

(Read '*The Sacred Numerology of Thoth*' to release yourself from that fraud)

Lastly, we should look at our system of mapping, or gridding this planet we inhabit.

A planet that is that 3rd rock from a sun that is part of a *binary* sun system revolving around a Galaxy called the Milky Way in this Universe that we are currently inhabiting.

Could our current system of cartography (read science of map-making) be flawed?

If so, is a Cartesian Co-ordinate system using Greenwich, England as its focal point another patently absurd artifice that is keeping us trapped in a Hu-man conditioned reality that has us locked in mediocrity?

And if so, what is going to set us FREE?

Maybe a novel or new way of viewing the Earth's grid system.

This book is an attempt to do just that.

So, let go of your current view, and try on something new, and you just might be surprised at what you see.

t. gyurme

CHAPTER ONE

Earth's Grids – Typography

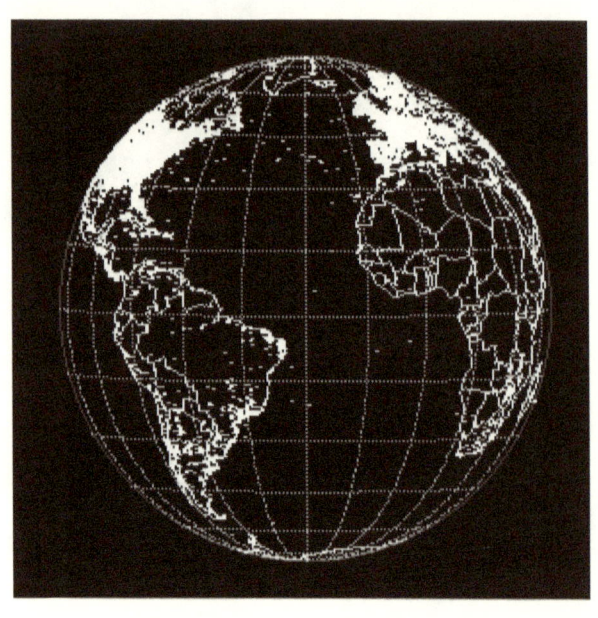

In typography, the grid system, also called grids, refers to the practice of laying out of a page using a grid of different-sized and -shaped rectangles to align items to. The purpose of the grid is to help the designer organize the content of the page in a pleasing and easy-to-digest manner for the viewer by imposing structure and cohesion to the layout.

The use of grids is quite common in graphic design and the concept is also used in interior design. Some allege that the idea of grids is similar to ideas of certain new-age versions of feng shui.

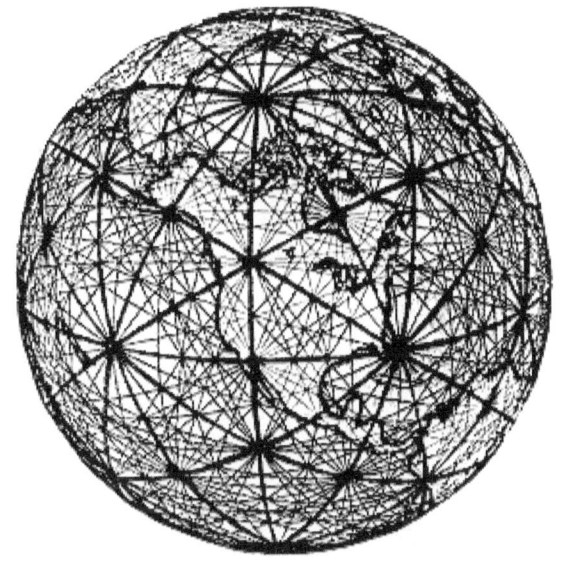

Planetary 'energy' grid theory falls under the heading of pseudoscience. It operates through geometrical patterns that follow a specific symmetry linked to sacred geometry.

The grids meet at various intersecting points forming a grid or matrix. This is equivalent to the acupressure points on our bodies. These grid points can be found at some of the strongest power places on the planet.

Plato, the Greek philosopher, who lived approximately 427-347 BC recognized the patterns of the grids. He devised a theory regarding his belief that the earth's basic structure evolved from a simple geometric shape to more complex geometric shapes. These shapes became known

as the platonic solids. They are: cube (4), tetrahedron (3), octahedron (8), dodecahedron (12), icosahedron(20).

Plato, in his *Timeaus* associated each shape with one of the elements, earth, fire, air, ether, and water. The earth's energy grid from the beginnings of its evolutionary course, has evolved through each of these shapes to what it is today. Each shape, superimposed, one upon the other to create a kind of all encompassing energy field that is the very basis of Earth holding it all together.

Curry Lines

Earth radiation is a theoretical geophysical phenomenon described primarily by the German authors Manfred Curry and Ernst Hartmann.

They both describe a mystic force field (similar to Odic force, Mana, Qi) that covers the earth at regular intervals and can be detected by dowsing using a divining rod. It is not supposed to be detectable by common scientific instruments but some still connect it to telluric currents,

which are actual phenomena, detectable by scientific instruments.

Placing people or other living things in certain spots of the earth radiation knots is believed to be beneficial/harmful depending on radiation flow direction. It connects to the Gaia philosophy and vitalist school and is very popular in certain New Age circles in Europe, especially in Germany.

The radiation is described as a grid-like arrangement with lines at regular distances:

- Curry lines are supposed to be circa 3 meters apart (with variations), diagonally to the poles, east to west.
- Hartmann lines run both east-west and north-south forming a grid across the earths surface with a distance of circa 2 meters in the north-south direction and 2.5 meters in the east-west direction.
- Ley lines are man-made energy lines, created by stone formations such as stone ships or other ancient archaeological structures. The knowledge of creating Ley lines is supposed to be lost.

Some proponents of this theory will also mention Schumann Waves/Resonance (which, like telluric current, is an actual scientifically detectable phenomena) and Black Lines.

The Schumann Resonance is a set of spectrum peaks in the ELF portion of the Earth's electromagnetic field spectrum. Schumann resonance is due to the space between the surface

14

of the Earth and the conductive ionosphere acting as a waveguide. The limited dimensions of the earth cause this waveguide to act as a resonant cavity for electromagnetic waves in the ELF band. The cavity is naturally excited by energy from lightning strikes. Since the seventh overtone lies at approximately 60 Hz, the cavity is also driven by the North American power grid.

The lowest-frequency (and highest-intensity) mode of the Schumann resonance is at a frequency of approximately 7.83 Hz. Detectable overtones extend upwards into the kilohertz range. The phenomenon is named after physicist Winfried Otto Schumann who predicted it mathematically in 1952, but it was first observed by Nikola Tesla and

formed the basis for his scheme for broadcast power and wireless communications (Tesla 1905). Continued

Ley lines refer to hypothetical alignments of a number of places of geographical interest, such as ancient monuments and megaliths. Their existence was suggested in 1921 by the amateur archaeologist Alfred Watkins, whose book *The Old Straight Track* brought the alignments to the attention of the wider public.

The existence of alignments between sites is easily demonstrated. However, the causes of these alignments are disputed. There are several major areas of interpretation:

- *Archaeological:* A new area of archaeological study, archaeogeodesy, examines geodesy as practiced in prehistoric time, and as evidenced by archaeological remains. One major aspect of modern geodesy is surveying. As interpreted by geodesy, the so-called

ley lines can be the product of ancient surveying, property markings, or commonly travelled pathways. Numerous societies, ancient and modern, employ straight lines between points of use; archaeologists have documented these traditions. Modern surveying also results in placement of constructs in lines on the landscape. It is reasonable to expect human constructs and activity areas to reflect human use of lines.

- *Cultural:* Many cultures use straight lines across the landscape. In South America, such lines often are directed towards mountain peaks; the Nazca lines are a famous example of lengthy lines made by ancient cultures. Straight lines connect ancient pyramids in Mexico; today, modern roads built on the ancient roads deviate around the massive pyramids. The Chaco culture of Northeastern New Mexico cut stairs into sandstone cliffs to facilitate keeping roads straight.
  - *New Age:* Some writers widely regarded as pseudoscientific have claimed that the ley lines and their intersection points resonate a special psychic

or magical energy. These theories often include elements such as geomancy, dowsing or UFOs.

Some similar believe these points on lines have electrical or magnetic forces associated with them.

- *Skeptical:* Skeptics of the actuality of ley lines often classify them as pseudoscience. Such skeptics tend to doubt that ley lines were planned or made by ancient cultures, and argue that apparent ley lines can be readily explained without resorting to extraordinary or pseudoscientific ideas.

Black Lines

Black lines seem to be naturally generated, although quite how is not known. They may be localized and do not form a network in the same way as Hartmann and Curry lines. This could be similar in nature to the "sha", or deadly energy lines of Chinese Feng-Shui. They can be curved, straight, at ground level or higher, even found in the upper levels of buildings. There have been described 2 types of Black lines, one as "black and depressed", the other as "shiny, black, hard and sharp." They could possibly represent the flow lines of a negative type of "orgone-type" energy as described by Wilhelm Reich.

Hartmann Net or Hartmann Lines

The Hartmann net consists of naturally occurring charged lines, running North-South and East-West. It is named after Dr. Ernst Hartmann, a well regarded German medical doctor, who first described it soon after the second world war. Alternate lines are usually positively and negatively charged, so where the lines intersect it is possible to have double positive charges and double negative charges, or one positive and one negative charge. It is the intersections that are seen to be a source of potential problems.

The Hartmann Net appears as a structure of radiations rising vertically from the ground like invisible, radioactive walls, each 21 centimetres (9 inches) wide. The grid is magnetically orientated, from North to South they are encountered at intervals of 2 metres (6 feet 6 inches), while from East to West they are 2.5 metres (8 feet) apart.

Between these geometric lines lies a neutral zone, an unperturbed micro-climate. This network penetrates everywhere, whether over open ground or through dwellings.

The Hartmann net has been defined using the Chinese terms of Yin and Yang. The Yin (North-South lines) is a cold energy which acts slowly, corresponds to winter, is related to cramps, humidity and all forms of rheumatism. The Yang (East-West lines) is a hot, dry rapidly acting energy. It is related to fire and is linked to inflammations.

The points formed by the intersection of these lines, whether positive or negative, are dynamic environments sensitive to the rhythms of the hours and the seasons.

It has been suggested that both the Curry grids and Hartmann Net are earthing grids for cosmic rays that constantly bombard the Earth, and that they can be distorted by other things, such as geological fault lines and underground mining. It is also possible to have spots where the Curry and Hartmann lines cross, causing further potential problems. These spots are generally seen to be more detrimental than a single crossing within the Curry or Hartmann system.

CHAPTER TWO

Becker-Hagens Grid Map

The husband-and-wife team of Becker and Hagens - created the map below. Becker is a Professor of Industrial Design at the University of Illinois, Chicago, and Bethe Hagens is a Professor of Anthropology at Governors State University.

Though this team may not seem to be the most likely pair to create such a globe re-fashioning grid system, remember that pure scientists are not a reliable source for understanding our complete Universe. (see '*I AM*' *In a Post-Modern World*, also available on my website, for more information on this topic)

So the fact that they Don't have a scientific pedigree, is what gives them credibility, at least to me. So, as the reader is going to interpret that which the will, in any event, please do yourself a favor and relax your critical mind while perceiving that which another has created.

The Becker-Hagens grid cracked the code of the Platonic Solids' positions on Earth.

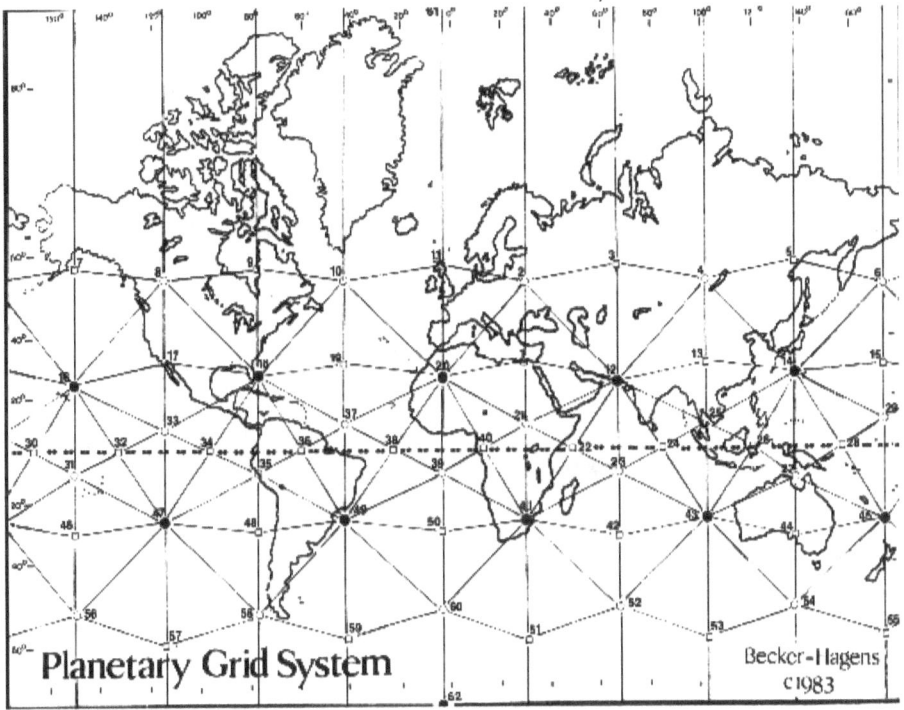

Planetary Grid System

Becker-Hagens
c1983

Becker and Hagens show us how these two shapes were first discovered, and then how they worked into the ultimate Earth Grid, which they called the Unified Vector

23

Geometry 120 Polyhedron, or the UVG 120 Earth Star.

They ascribe this discovery to the work of Ivan P. Sanderson, who was the first to make a case for the structure of the icosahedron at work in the Earth. He did this by locating what he referred to as *vile vortices,* or areas of the Earth where mysterious disappearances, mechanical failures and time-space distortions were seen to occur.

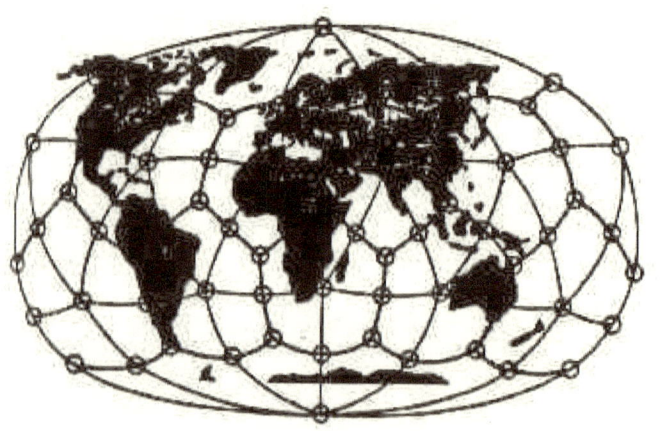

These 3 scientists worked off of their combined talents to determine this "matrix of cosmic energy" that encircles the Earth. With this new formulation of the Global Grid, Becker and Hagens write:

These new lines and points, in conjunction with Sanderson's, now matched most of the earth's seismic fracture zones and ocean ridge lines as well as outlined worldwide atmospheric highs and lows, paths of migratory animals, gravitational anomalies, and even the sites of ancient cities.

Becker and Hagens' attention was drawn to this research through the work of Chris Bird, who wrote his article on the "Planetary Grid" in the New Age Journal of May 1975. They were so overwhelmed with this new information that they eventually set up a meeting with him to discuss the work. Soon after this, they "completed" the Grid, making it compatible with all the Platonic Solids, by inserting a creation from Buckminster Fuller's work. Quote from their work...

We propose that the planetary grid map outlined by the Russian team Goncharov, Morozov and Makarov is essentially correct, with its overall organization anchored to the north and south axial poles and the Great Pyramid at Gizeh. The Russian map, however, lacks completeness, in our opinion, which can be accomplished by the overlaying of a complex, icosahedrally-derived, spherical polyhedron developed by R. Buckminster Fuller. In his book *Synergetics 2*, he called it the "Composite of Primary and Secondary Icosahedron Great Circle Sets." We have shortened that to Unified Vector Geometry (UVG) 120 spheres. We use the number 120 due to its easy comprehension as a spherical polyhedron with 120 identical triangles - all approximately 30, 60 and 90 in composition.

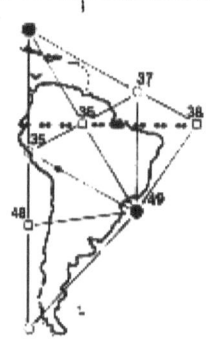

South America's grid triangle forms the continent around itself. In looking at the southernmost tip of South America, you can see how the force of node number 58 pushes the land away from its due south trend and towards the east. Then, if you look at node 49, on the middle of the East Coast of South America near Rio de Janeiro, you can clearly see how the force of the node has pushed the landmass of the continent into a rounded shape.

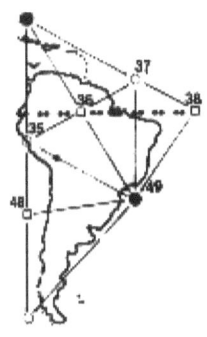

Looking at Australia, you can clearly see that the whole continent, and especially the northwest side, forms very precisely within the stretching forces of this area of the Global Grid. Here, if we look to the exact north and middle of Australia on point 27, we see a circular "node point" displacing the land around it and forming the Gulf of Carpenteria. Again, the nodes themselves have shaped the land into circular "vortices," repelling the continental mass from themselves and in this case, also shaping the outline of the island directly above Australia.

Node 44 is precisely aligned with the bottom of Antarctica, and we see either edge "drooping" to the right or the left from this. According to Richard Lefors Clark, this is the "bowtie" energetic configuration showing itself in the Australian landmass, which he calls a "diamagnetic energy vortex." We will examine how such a shape could be formed by the curved, spiraling energies that make up the grid below. Clark also shows us that the Gulf of Mexico follows this same curving energetic organization, as well as the two coastlines of the continental United States.

Africa shows the combined action of two larger triangles, one with the point facing downwards on the African continent and its neighbor pointing upwards, griding the Indian Sea.

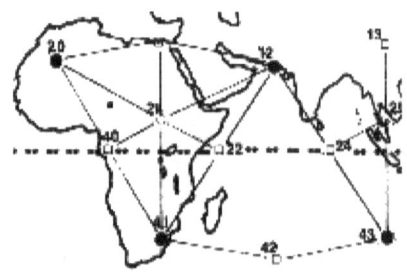

The downward-pointing triangle certainly describes Africa well, and the extra land on the western side can be explained by the pressure coming from the Mid-Atlantic Ridge, which is almost precisely the same as the Atlantic grid line formed by points 10, 19, 37, 38, 39 and 50. The east coast of Africa and Saudi Arabia follow the eastern side of the triangle very nicely, on the grid line from point 41 to point 12.

Node point 22 gently pushes in on Africa's east coast, causing it to dip inwards as it travels and forming the

Somalia Basin. We can also assume that the "cracks" separating Africa and Saudi Arabia were caused by the expansion of the Grid, forcefully pulling the land into alignment. The "crack" to the immediate left of point 12 follows the grid line exactly. And finally, the upward-pointing triangle can be clearly seen to cut its way right into the land, with Africa's coast defining its left side and India's coast defining its right.

Node point 22 gently pushes in on Africa's east coast, causing it to dip inwards as it travels and forming the Somalia Basin. We can also assume that the "cracks" separating Africa and Saudi Arabia were caused by the expansion of the Grid, forcefully pulling the land into alignment. The "crack" to the immediate left of point 12 follows the grid line exactly. And finally, the upward-pointing triangle can be clearly seen to cut its way right into the land, with Africa's coast defining its left side and India's coast defining its right.

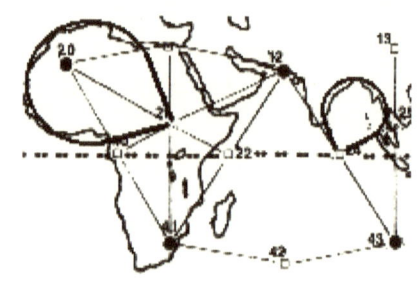

In this next illustration, we can see how point 21, in the center of the African continent, works with its above-right partner point 20 in providing a framework for the bow-shaped energy vortex that shaped the northeast coast of Africa.

Point 20 is the only "vile vortex" that is significantly inland; 41, near South Africa, and 42, near India, are the only others that touch land at all! This is probably due to the incredible strength that they possess, which seems to repel landmasses. We can see how the northeast coast of Africa is indeed very circular, and point 20 is directly at the center of this "bowtie" of energy.

We see Dr. Lefors Clark's bow-shaped "diamagnetic energy vortex" shaping the land in a smooth curve. We can also see the same curving formation created by the east coast of India and the west coast of the China / Korea / Vietnam area of Asia. This vortex is balanced between points 24 and 13. We remind ourselves that these smooth curves illustrate the spiraling nature of the superstrings that make up these geometric energy fields, here expressed as spiraling lines of magnetic force.

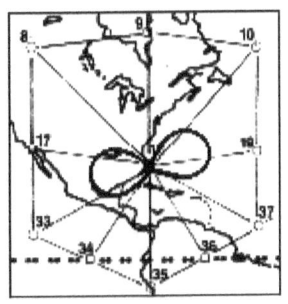

According to Dr. Clark, the point at the center of the "bowtie" becomes a magnetic null zone of zero gravity when appropriately triggered by the right geometric positions of the Sun, Moon and Planets to the Earth. In those moments of conjunction, there is a "hyperdimensional bleed-through," and we then get time dilations.

The next three pictures, showing the circular grid energy formations surrounding the east coast of China and surrounding area:

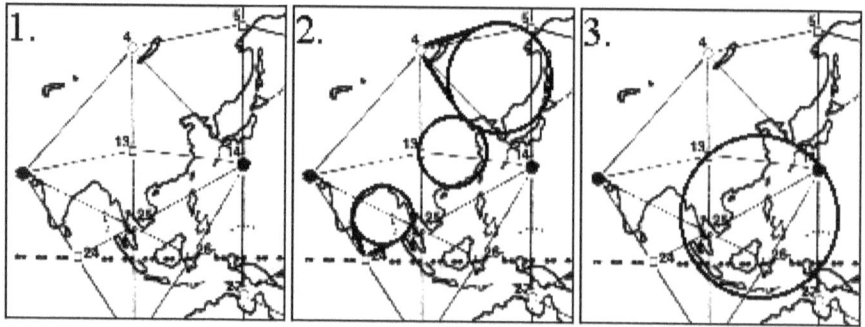

In picture number 1, we have our straight grid map of China for comparison. Picture 2 has three circular vortexes in it. We have already discussed the bottom left vortex in India and China. Then, the center vortex shapes the East Coast of China, based off of and emanating from point 13. The uppermost and largest vortex in Picture 2 shapes the entire Japanese and Russian archipelago, and is centered in and emanating from point 4.

In picture number 3 we have a vortex centered in Sanderson's "vile vortex" near Japan, the "Devil's Sea," here listed as point 14. This vortex is "equalized" by points 25 and 26, both of which are equidistant from the edge of the circle. And obviously, Indonesia precisely traces the bottom left of the circle itself. This is obviously a very strong vortex to be able to shape the land of Indonesia from where it stands, according to Becker and Hagens.

Another area of continuing disappearances and mysterious time-warps is the Devil's Sea located east of Japan between Iwo Jima and Marcus Island. Here events have become so sinister that the Japanese government has officially designated the area a danger zone. This area was significant enough to the work of Bermuda Triangle author Charles Berlitz that he wrote a whole book dedicated to it and the "bigger picture," entitled The Devil's Triangle. It is becoming more and more clear that our geometric shapes, expressed as the consciousness unit expanded to a planetary scale, are far more than abstract concepts of theoretical physics. What we have here are direct, quantifiable and measurable phenomena, and these geometry-based Grid patterns are simply the simplest, and therefore the best solution to the problem.

The "grid bands" on the Earth and how their effects shaped the Ring of Fire', the flowing of the Nile River, the "node point" of the Egyptian northern coast centered in 'Giza, and the vertical structure of the Yucatan Peninsula. Now, with the full map of the Becker-Hagens grid, we can see a great deal more from the effect of these straight longitudinal lines.

By referring back to the main map, the reader can clearly see that the entire Eastern side of Hudson Bay in Canada precisely follows the line from 18 in Florida to 9 in the Bay to 61 at the North Pole. Furthermore, all of England is precisely within the line created by point 20 in Africa, through point 11 in England to point 61 at the North Pole. So, there are a variety of ways to see this energy at work on Earth. One can begin to visualize this Grid energy as a living net of "wires" that are tightly stretched over a thin balloon. It is obvious to us that what we think of as randomly placed continents are actually conforming to this massive energy, disappearances, gravity loss, levitation and other related phenomena.

Becker-Hagens explain how these grid points seem to attract large population centers.

Look at the South American landmass. Not only does it fit perfectly a Grid Triangle, but we can see a circular, bowtie-shaped energy at work in the actual shaping of the landmass itself.

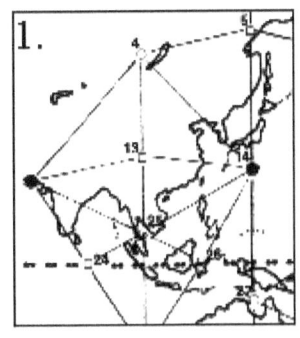

This "bowtie" is nearly centered within the diamond that is formed between points 18, 35, 37 and 49. We already saw the African "vile vortex" shaping a similarly large "bowtie" in the above diagrams. The offset of the South American landmass from being precisely within this "diamond" again could be accounted for by the pushing of the Mid-Atlantic Ridge, which follows the lines of the Global Grid with amazing precision.

Looking back at the Becker-Hagens map, it is quite easy for us to see how this line clearly demarcates the separation between the continents, just as the Mid-Atlantic Ridge is the point of expansion between the two continental plates.

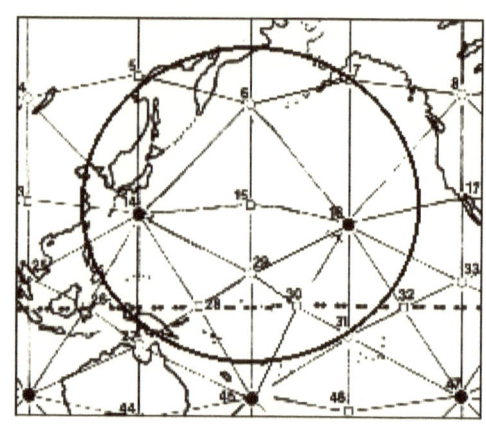

An elliptical-shaped gravity field is clearly visible when the center is placed directly in node 15. If we use any standard image editing program and "drag out" an ellipse using point 15 as a center, we can indeed align it precisely with the island formations to the far west of our diagram. We can see the amazing connections of this energy formation as soon as we start looking at the diagram. We can try other center points besides 15, but the ellipse will not fit anywhere else as precisely as it does right at that spot.

This massive energy vortex seems to provide the clearest Grid counterpart for the existence of the Ring of Fire, which is a ring of volcanoes and tectonic activity surrounding the Pacific Ocean. When we look at this "grid ring" carefully, we can see that it represents the perfect fusion between the Earth's landmasses and the Global Grid. Going clockwise from the 12:00 point, the ring will

36

perfectly touch a "square" of grid points, as 7, 31, 27 and 5. (Node point 27, near Australia, is the only one that isn't touched exactly.

We can also see that this ellipse is well defined by points 14 and 16, again Sanderson's incredibly powerful "vile vortices," the points of the icosahedron. We have already seen how the incredible gravitational force of these "vortices" was able to shape the entire upper Western half of Africa into an elegant, circular shape. Now, we are seeing two of these vortices working together to form an even larger shape. The ring forms part of the East Coast of China, as well as a good part of the upper Russian coastline surrounding node 5. It also defines part of the southern coastline of Alaska.

The grid points 14 and 16 would be akin to the two poles of the dividing cell. The grid lines provide us with a simplified depiction of the "spindle fibers" that form in cell mitosis. The actual ring of energy that is created forms a precise analog of the nuclear membrane of the cell, as it continues its expanding, elliptical process of division.

CHAPTER THREE

Vile Vortices

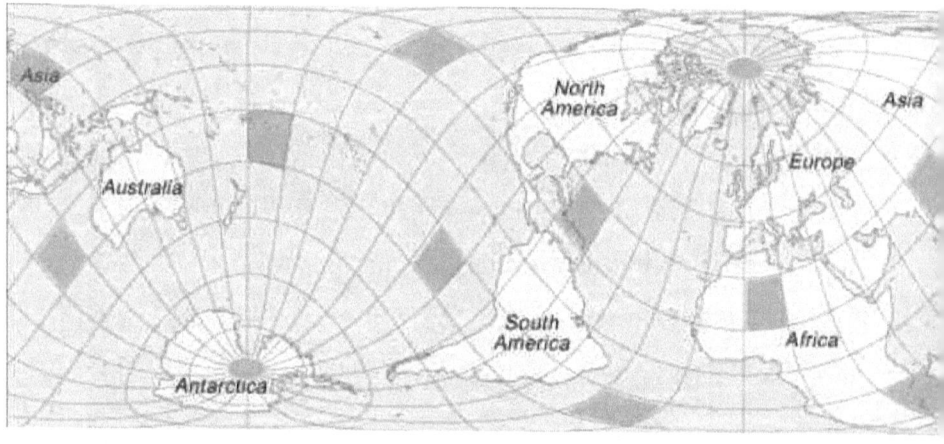

Vile Vortices refer to a claim that there are twelve geometrically distributed geographic areas that are alleged to have the same mysterious qualities popularly associated with the Bermuda Triangle. The best-known of these are the Bermuda Triangle itself, the Devil's Sea near Japan and the South Atlantic Anomaly.

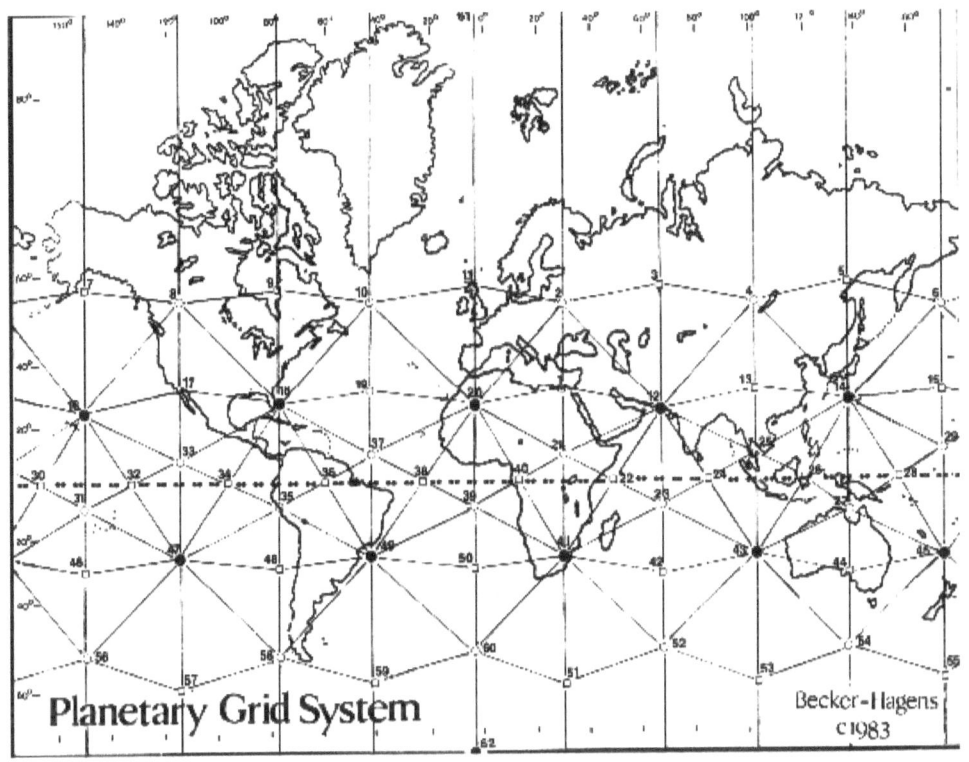

Merging the Vile Vortices into the Planetary Grid

Vile Vortices

Definition: 1. vile –|t. plural of vili meaning miserable. 2. vortices –|t. plural of vortici meaning whirl or vortex.

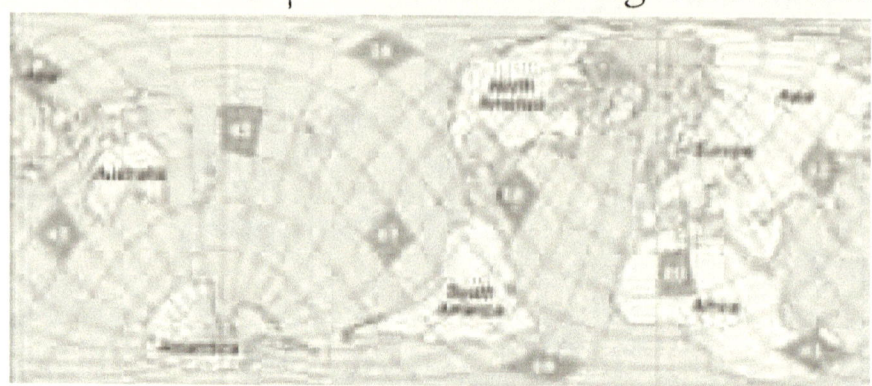

Explaining the Vile Vortices

By definition, the Vile Vortices would be miserable whirlers but actually they are twelve vertex points of a planetary grid (see Figure 1) originally plotted by Ivan T. Sanderson, a naturalist and paranormal investigator. Sanderson first coined the term, "Vile Vortices"1 in his article "The Twelve Devil's Graveyards Around the World" (Saga magazine, 1972).

The best-known Vile Vortices are the Bermuda Triangle, the Dragon's Triangle (Devil's Sea), and the South Atlantic Anomaly. However, each of these twelve geographic areas is credited with instances of magnetic anomalies and other unexplained phenomena.

Plotting the Vile Vortices

Ivan T. Sanderson, the founder of the Society for the Investigation of the Unexplained, was avidly interested in investigating ship and plane disappearances linked to the paranormal. In the late 60's he focused his attention on ten areas that were approximately equidistant and were the subjects of reported unexplained incidents and/or electro-magnetic distortions[2].

Ten of Sanderson's Vile Vortices are located in the earth's tropical climates; five of them fall within the Tropic of Cancer and the other five within the Tropic of Capricorn. The remaining two Vile Vortices are located at the North and South Poles. Together the Vile Vortices form the vertices of an icosahedron (a 20-faced polyhedron) [3].

Sanderson theorized that hot and cold air and sea currents crossing these lozenge-shaped areas might create the electromagnetic anomalies responsible for the disappearances of planes and sea-going vessels and the reported mechanical and instrument malfunctions in these areas. 1

Merging the Vile Vortices into the Planetary Grid

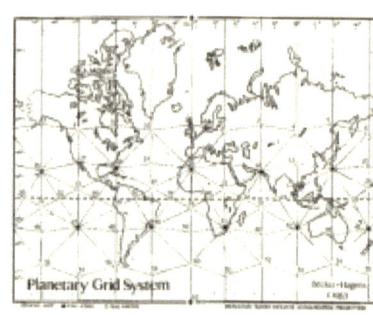

Figure 2 -

In 1973, three Soviet scientists (Nikolai Goncharov, Vyacheslav Morochov, and Valery Makarov) extended Sanderson's theory in their article, "Is the Earth a Large Crystal?" published in a Soviet science magazine.3 They postulated that "a matrix of cosmic energy" made up of twelve pentagonal plates covers the earth. The article claimed that the junctions of any three of these plates (62 junctions in all) possessed unusual properties such as advanced prehistoric cultures, unique wildlife, or other mysterious phenomena.

The work of the Russian scientists outlined a planetary grid that built on Sanderson's original twelve Vile Vortices (see figure 2), overlapping Sanderson's icosahedron with a combination of icosahedrons and dodecahedrons in parallel with many of the earth's seismic fracture zones, ocean ridges and additionally portrayed the earth's atmospheric highs and lows, routes used by migratory animals, and gravitational anomalies as well as the locations of ancient civilizations."2

However, planetary grid systems date back to the time of Plato (approx. 427-347 BCE). Plato theorized that the earth's basic structure evolved from simple to complex geometric shapes; the cube, tetrahedron (3 sides), the octahedron (8), the icosahedron (20), and dodecahedron (12), known today as the platonic solids. He associated four of the five shapes with, respectively, the elements of earth, fire, air, and water and the dodecahedron with the earth's prana/aether (life sustaining force).2,4

David Hatcher Childress, author of *Mapping the World Grid* observes "...we are speaking about an intelligent geometric pattern into which, theoretically, the Earth and its energies are organized—and possibly in which the ubiquitous megalithic sites are also positioned."5

The Vile Vortices on the Planetary Grid

Vortices in the Tropic of Cancer	Vortices in the Tropic of Capricorn
12. Mohenjo Daro, site of the Rama Empire 14. The Devil's Sea (Dragon's Triangle) 16. Hamakulia, near Hawaii, the scene of high volcanic activity 18. The Bermuda Triangle	41. The Zimbabwe Megaliths 43. Wharton Basin, the site of the Wallaby Fracture Zone, 45. The edge of the Hebrides Trench near the Fiji Islands

20. The Algerian megalithic ruins 61. The North Pole	47. Easter Island Colossi 49. The South Atlantic Anomaly 62. The South Pole

Related Categories:

Bermuda Triangle | Dragon's Triangle | South Atlantic Anomaly | Bridgewater Triangle | Bennington Triangle

Resources:

Figure 1: Public Domain Image with numbering enhanced for easier reading

Figure 2: The Planetary Grid System shown in figure 2 was inspired by an original article by Christopher Bird, "Planetary Grid," (New Age Journal #5, May 1975, pp. 36-41). The hexakis icosahedron grid, coordinate calculations, and point classification system are the original research of Bethe Hagens and William S. Becker. This map is displayed with permission of the authors in cooperation with Governors State University, University Park, Illinois.

1. Dan Shaw. "12 Devil's Triangles-10 Vile Vortices Around the World." Vortex Maps 2007. 27 Apr 2008

<http://www.vortexmaps.com/devils-triangle.php>.
2. William Becker, Beth Hagens. "The Planetary Grid–A New Synthesis." Bibliotecapleyades Unknown. 28 Apr 2008

3. "Vile Vortices." Wikipedia 21 Feb 2008. 18 Apr 2008
4. David Hatcher Childress. "Mapping The World Grid." Bibliotecapleyades unknown . 28 Apr 2008

CHAPTER FOUR

The Bruce Cathie Grid System

Back in the 1950's, an airline Pilot in New Zealand noticed an anomaly in extraterrestrial sightings.

From that he hypothesized a grid system of ley lines and power lines encompassing the earth's surface.

The following is a synopsis of these grids.

Forty years ago, Captain Cathie began his investigation into the World Grid® system. His initial interest was sparked by a sighting of an unidentified flying object over one of the local harbours in Auckland, New Zealand. He was immediately aware that there were machines, other than normally-known aircraft, moving through our airspace, which appeared to be more highly advanced technically than our own.

As time went on, it became known to him that many other fellow pilots and radar operators around New Zealand were also sighting similar machines which were carrying out manouvres and flying at speeds that exceeded the capabilities of all known types of aircraft. It became obvious that either our own scientists had very advanced secret knowledge or we were possibly being visited by highly advanced beings from outer space. Later, as more information was received, it appeared that both these answers were possible.

The captain commenced his own research project and plotted sighting positions onto an airways map which covered the New Zealand area and eventually discovered that a definite pattern emerged which indicated that the unknowns were under intelligent control. Several years later a complex geometric grid system was completed which covered the whole world surface. The patterns were eventually broken down into mathematical and geometric coordinates which could then be directly related to gravity, light speeds and the earth's magnetic fields. From that point on the research developed at a faster pace and culminated in a series of unified equations and sets of unified tables which could be related to all branches of scientific research.

A major breakthrough occurred when it was discovered that the mathematical base, which was harmonic in nature, could be related to the detonation of atomic devices. This fact indicated that all atomic activity was geometric in nature and that an all out atomic war was illogical. All detonation times and places could be pre-calculated by both sides in this type of war which virtually meant stalemate. It was after this fact was made known, that the research activity of Captain Cathie became noticed by certain governments and intelligence agencies. Tests were made of the knowledge held and the harmonic mathematical theories were then admitted to being correct.

The research is ongoing and now, many years later, a computer program has been produced which carries out all the required calculations related to the light, gravity and

49

earth magnetic field values. It has been found that many ancient building sites such as Stonehenge, the Great Pyramid, etc., show relationships with this unified harmonic mathematical system and indicates that the knowledge is not new. The civilizations at these times must have had scientific knowledge of a highly advanced nature.

We are now in the process of relearning this lost science. If the discovered harmonic unified equations can eventually be accepted by the scientific community as correct, then we may be on the threshold of true space travel which has no barriers. The equations show that light speed is not a constant and that time itself is a variable. Under these conditions there is no place out there in space that we cannot reach. Obviously, there is more mathematical knowledge to be discovered, but it appears that the code has been broken. Soon, with a bit of luck, we on Earth may have the chance to join our neighbours out there in the vast blackness of space.

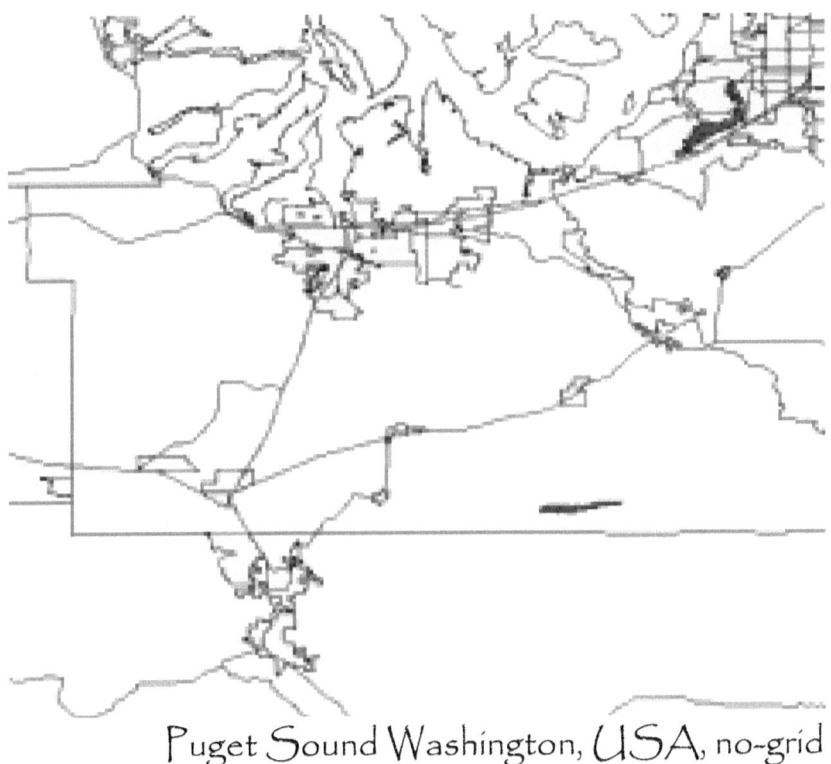

Puget Sound Washington, USA, no-grid

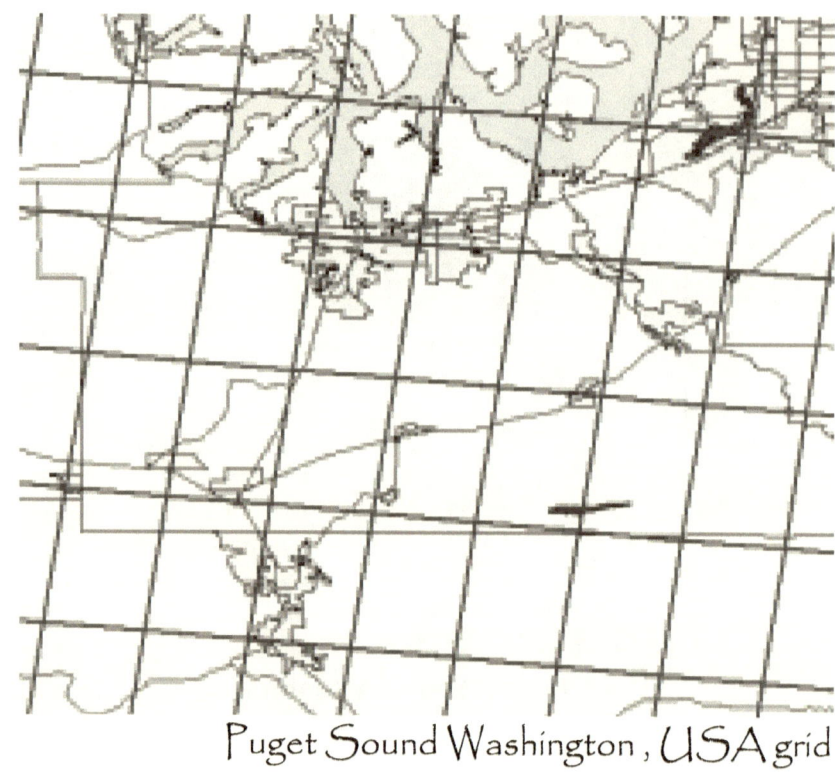

Puget Sound Washington, USA grid

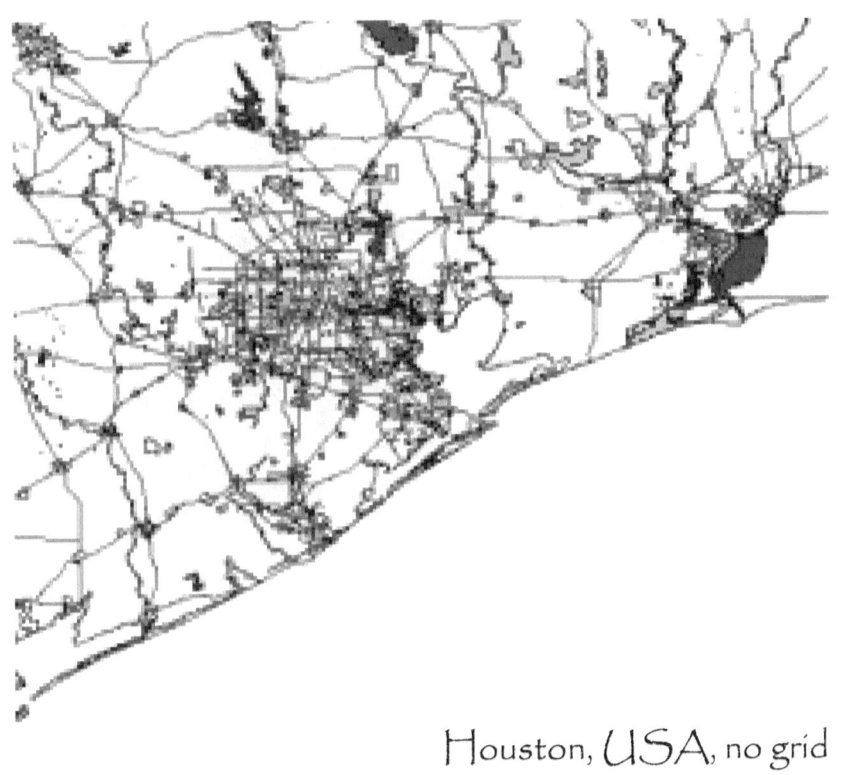

Houston, USA, no grid

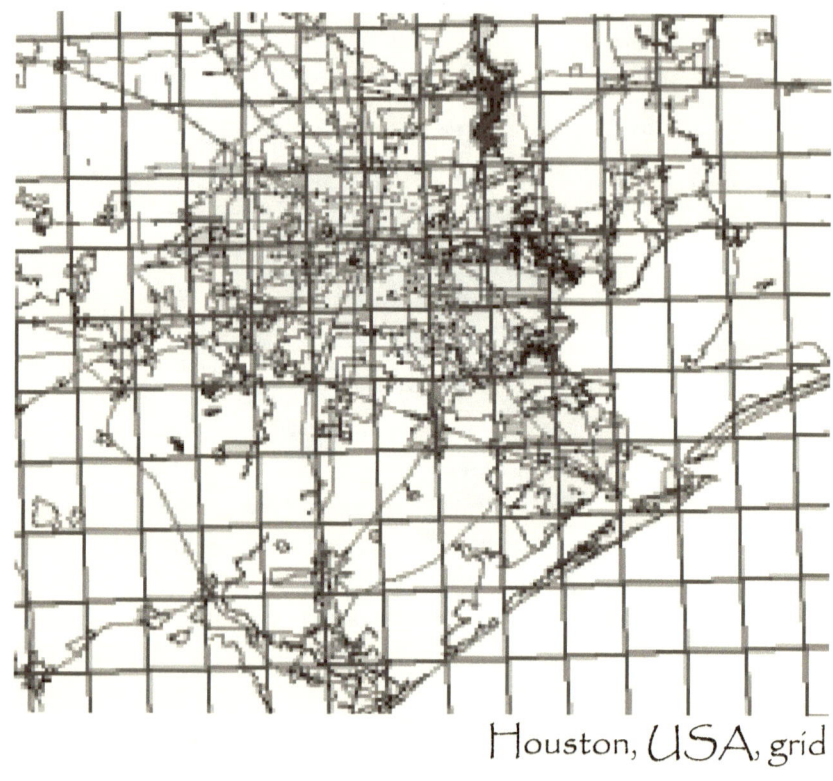

Houston, USA, grid

Aukland, N. Island, grid

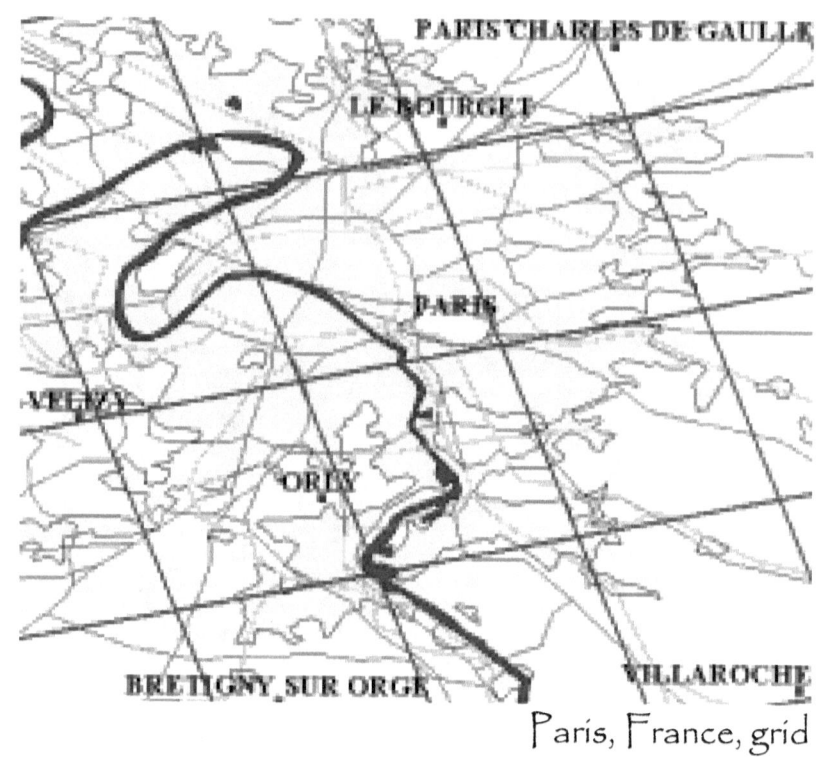

Paris, France, grid

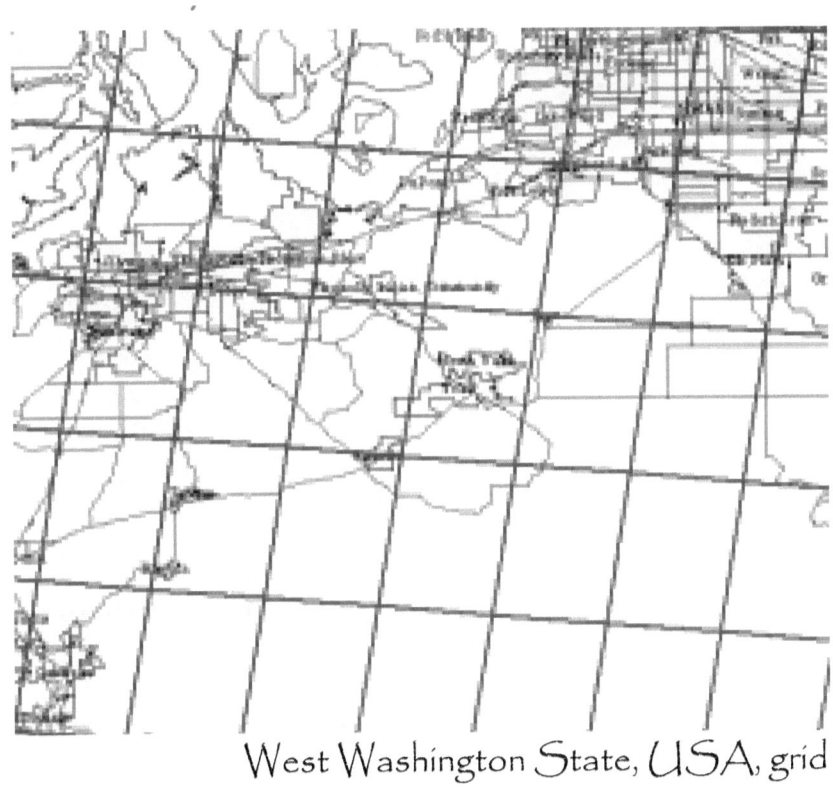

West Washington State, USA, grid

CHAPTER FIVE

Grid Harmonic Evaluation For The Destruction of the Twin City Bridge: by Bruce L. Raphael (C) August 2007

MAP

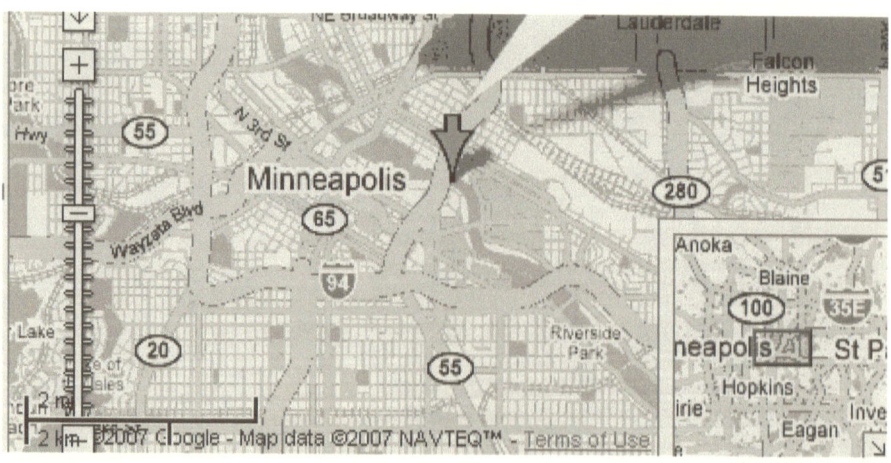

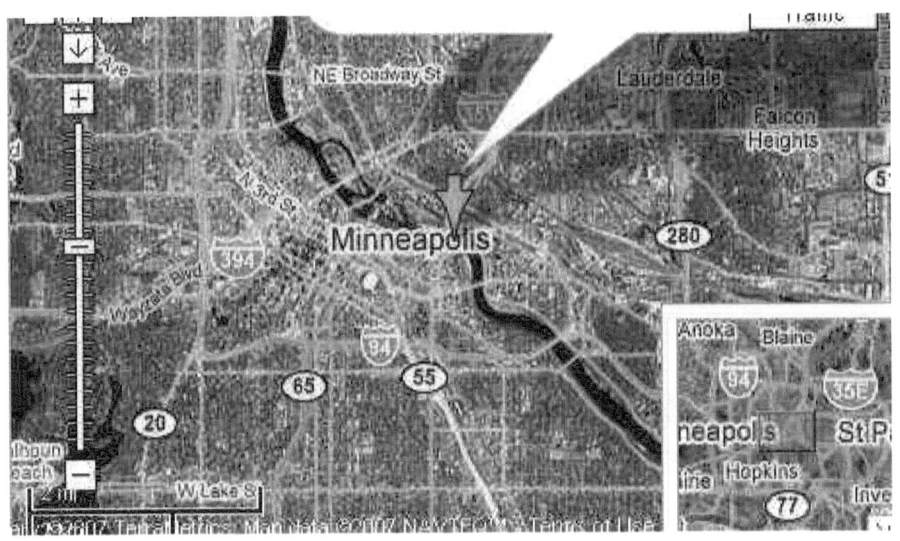

From Sorcha Faal (seen at GLP August 4, 2007)

August 02, 2007 Reports from Russia"s Institute of Solar-Terrestrial Physics located in Irkutsk are reporting today that their Siberian Solar Radio Telescope (SSRT) detected a "massive" ultra low frequency (ULF) "blast" emanating from Latitude: 45° 00' North Longitude: 93° 15' West at the "exact" moment, and location, of a catastrophic collapse of a nearly 2,000 foot long bridge in Minneapolis, Minnesota.

To the horrific destruction of the Interstate 35W Bridge which spanned the Mississippi River we can read as reported by the Star Tribune News Service:

"The 1,907-foot bridge fell into the Mississippi River and onto roadways below. The span was packed with rush hour traffic, and dozens of vehicles fell with the bridge leaving scores of dazed commuters scrambling for their lives.

Nine people were confirmed dead as of 4 a.m. today. Sixty were taken to hospitals and 20 people were still missing this morning. Authorities said they expected the death toll to rise."

These reports further state that one of the United States primary research organizations into acoustic weapons research is Augsburg College, and which is located in Minneapolis, Minnesota, and most importantly less than 1 mile from the Interstate 35W Bridge collapse.

Clues were emerging on the possibility that an unusual non-terrorist assessment of the bridge collapse could point to unusual 3rd generation scalar technology. Scalar science is based on harmonic frequencies and geodetic math.

This analysis provides evidence for an hypothesis for the

destruction of the I35 Bridge in Twin Cities and leaves a clue to the recent take-over of the North Pole by the Russians' to control the Cathian Lunar Energy Grid and tap into more profound uses of Tesla technologies.

Study into Grid Harmonics and the Speed of Light at the Earth's surface is based on the math evidence below I calculated using my experience in energy grid math taught to me by Bruce Cathie himself in 1993.

See: Bruce Cathie maps the World Grid® with Gridpoint Atlas

To me I am astounded by the accepted statistical variance in the results. Too high to be coincidental.

Distance from Bridge impact to Augsburg College (a suggested place for an acoustic weapon device) is 1.3329 nautical miles N.B. 44.978333 N 93.24333 W Bridge location and Auburg location is 45.0 N 93.25 W

A) 1.3329 x 60 x 60 x 60 equals Harmonic 288 (twice the

speed of light harmonic: basic calculation: divide the grid spacing of 7.5 moas by the number of minutes in a circle 21600 = 288)

B) 1.3329 x 60 equals 79.974 divided by 143909.06 (speed of light in grid values equals harmonic 555 written as 5.5557...(harmonic 555 has been demonstrated to be the harmonic for Mother Mary apparitions based on the Castlemaine Australia distance to the Vatican)

C) 1.3329 x 6080 (grid feet per nautical mile) equals 8104.032 divided by 6 x 2 equals 27 Harmonic for Grid time of 27 hours per grid day.

D) Log of 113.29 equals 2.05419 twice the distance from grid pole A at the north pole to Geographic Pole.::::: Grid value distance is 1054 degrees

E) 1.3329 divided by 60 equals harmonic 222 as 0.022215 or 0.02222

F) 222.2 harmonic as written in minutes of arc is derived

from the ENERGY HARMONIC 2693645
where The Geometric harmonic Unified Equation at the
earth's surface

Reciprocal of harmonic 2693645 is 37124416
371244 x 60 equals 222(7440)

G) The Log of 1.3329 equals .12479 Harmonic 12.479.
Derived from the HALF VALUE of the Speed of
Light reciprocal Harmonic 3477 of 695 written as
69453074 (N.B. 143795.777 speed of light harmonic at
the Earth's surface SO 360 - 12.479 (stated harmonic in
the table is 12.7945) = 347.52

1.3329 Nautical miles equals 2.4685308 kilometres or
1.5329 miles

H) 68.68 divided by 1.3329 equals 51.51 angle measure
of the Giza Pyramid !! the value 68.68 nautical miles is
derived from the archaeological site at Mossman
Queensland.

See: http://geocities.com/worldtree_2000/diagrams.htm
! Ley-line bearing 180 degrees E to B

I) 1.3329 x 6080 x PI equals 2545: 2545 is the half the Polar Grid diagonal harmonic. It is also the square root harmonic of the temperature scale harmonic 648. 6477 expresses the sum of subatomic scale of elements.

BLR 8/5/07

Rush Allen comments:

http://www.eyeofsiloam.com/135W%20Collapse/index.html

"A 'massive' ultra low frequency (ULF) 'blast' could cause great loads to shift nominal tensile forces so as to "trigger" a failure event. But, there is just as much probability that there was more energy in the falling bridge and that the bridge triggered the ULF.

The fact that they are "simultaneous" suggests that the ULF is a side effect of Cosmic Will. It would be the sound made when a boxer lands an upper cut. The energy in the sound is nothing compared to the energy in the upper cut, but the sound wave carries the "message of stress" to much greater distances. The ULF was also found in the crash of a Jumbo Jet off New England.

CHAPTER SIX

STONE CIRCLES, CRYSTAL GRIDS, and PARTICLE ACCELERATORS: the PHYSICS OF PI (ARTICLE IN PROGRESS)

STONEHENGE--ANCIENT MEGALITHIC COMPUTER or a relic of ANCIENT ADVANCED TECHNOLOGY(OR BOTH)???

Is the geometric arrangement of stones at STONEHENGE only for use as an

ASTRONOMICAL CLOCK or does it have some
additional purpose?

The main circle of large stones is referred to by scientists as the SARSEN CIRCLE. Supposedly the SARSEN CIRCLE was once composed of 30 upright stones that were topped by 30 LINTEL'S that were mortised at each end. A lintel is referred to as a horizontal block that lies across(perpendicular) the upright stones. Of the 30 upright stones that were originally claimed to have been there only 24 are left standing in their original positions. Of the 30 LINTEL'S that lie across the tops of the upright stones only 5 remain in place. There is another circle outside of the SARSEN CIRCLE referred to as the AUBREY CIRCLE. What remains of the AUBREY CIRCLE is a series of 56 evenly spaced holes which vary in depth from two to four feet. Between the AUBREY HOLES and the SARSEN CIRCLE are two other circular patterns of holes named the Y HOLES and the Z HOLES. There are 30 holes in the Y series forming a circle roughly 35 feet outside the SARSEN CIRCLE, and supposedly only 29 holes in the Z series of holes which are about 5 to 15 feet outside of the SARSEN CIRCLE. Apparently the holes are not regularly spaced in the Y and Z holes as reported by some sources and each hole also has been found to contain a single BLUE-STONE fragment of the RHYOLITE variety of stone.

GEOMETRIC PATTERN of QUARTZ STONE.

It has been reported by ONE investigator into the STONEHENGE phenomena that STONEHENGE was apparently constructed to the universal geometric harmonics of LIGHT. This investigator states that the radius of the AUBREY CIRCLE is related directly to the SPEED of LIGHT; and the radius of the SARSEN CIRCLE corresponds to the square of the reciprocal of the speed of light.

It has also been reported that the larger STONES seem to have a CRYSTALLINE STRUCTURE and apparently may emit some form of electromagnetic radiation.

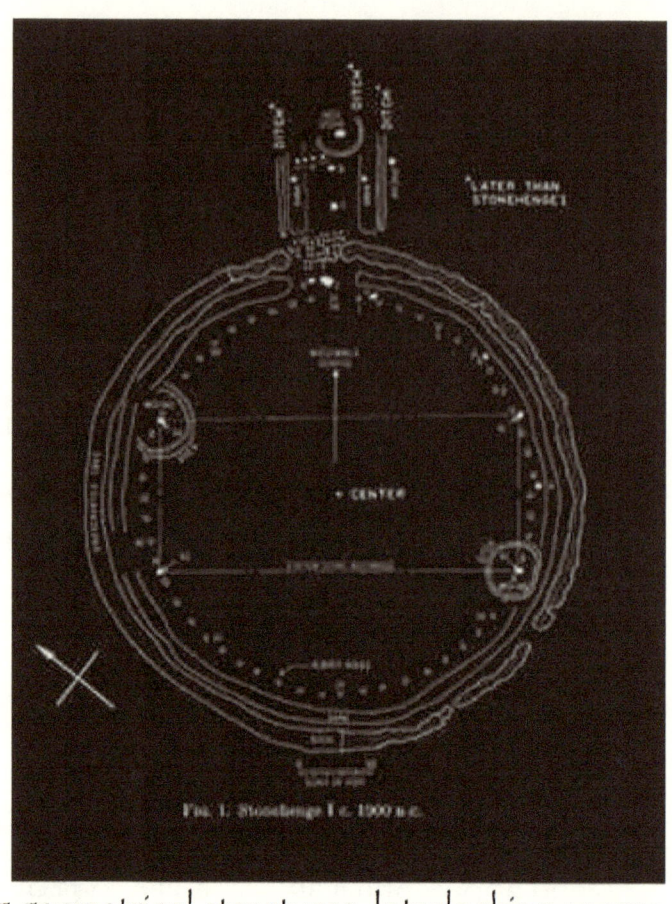

Fig. 1. Stonehenge I c. 1900 b.c.

Interlocking geometrical structures. Interlocking energy fields. Higher order energy patterns.

CRYSTAL(STONE)GRID. When a specific number of crystalline structures are aligned together in a highly specific geometrical pattern you create a CRYSTAL GRID. A Grid of structures. A set of crystalline structures. These grids or configurations can be used under the correct circumstances to generate energy fields that possess specific geometrical forms.

One such configuration is called the STAR of DAVID configuration pattern. In this configuration 6 crystalline structures of similar size are used. The structures are arranged in a pattern(within a circle) so that 6 of them are placed at the points of a hexagon with the termination(points) pointed inward. It has been said by one source that an ETHERIC crystal is supposedly being created by aligning SIX CRYSTALS in a STAR of DAVID configuration pattern. In the center of this STAR of DAVID crystalline grid structure it has been said by some individuals that the energy fields converge into the center to form a pillar(tower) of light. The STAR of DAVID configuration pattern is only one of many possible geometrical arrangements or forms that can be constructed with crystalline or stone structures. Each configuration of structures will produce different energy patterns or energy fields with different geometrical forms that will be dependent on the particular configuration that is constructed.

STONEHENGE possible being used as device to set up SPHERICAL energy fields.

Harmonic relationships that have been discovered at the STONEHENGE STRUCTURE would set up opposition to the LIGHT FIELDS at the earth's surface or those LIGHT FIELDS that would prevail in space. SPACE and TIME could be altered by the manipulation of the outer all enclosing field set up by the STONEHENGE CRYSTALLINE STRUCTURE.

The layout of STONEHENGE could be used as a plan or blueprint for a gigantic machine that could provide access for travel in space and even perhaps in time. Maybe STONEHENGE was once a device for manipulating energy fields located in our space-time continuum in order to somehow interfere with energy fields located in a different spacetime continuum (separate universe) or it could have been used for transporting matter(people) to and from far way spaces or even other universes. The very pattern of STONEHENGE is suggestive of the DISC SHAPED VEHICLES that could travel the skies.

STONEHENGE as model for a WHEEL (possibly a spinning wheel of light)--perhaps the complete device that was once operational could have even been a VORTEX GENERATOR.

STONEHENGE is a circle and some of the stones with the lintels on top are suggestive of the symbol PI which is the expression or symbol used by ancient greeks for the relationship of the circumference of a circle to it's diameter. It is also the sixteenth letter of the GREEK ALPHABET. It also seems apparent that a STAR of DAVID configuration can fit inside the STONEHENGE CIRCLE and within that there is a HEXAGON. I have also heard from certain sources that the builders of STONEHENGE used a BASE 6 number system instead of BASE 10 number system as we do today. Supposedly these sources claim that STONEHENGE is based on a BASE 6 number system.

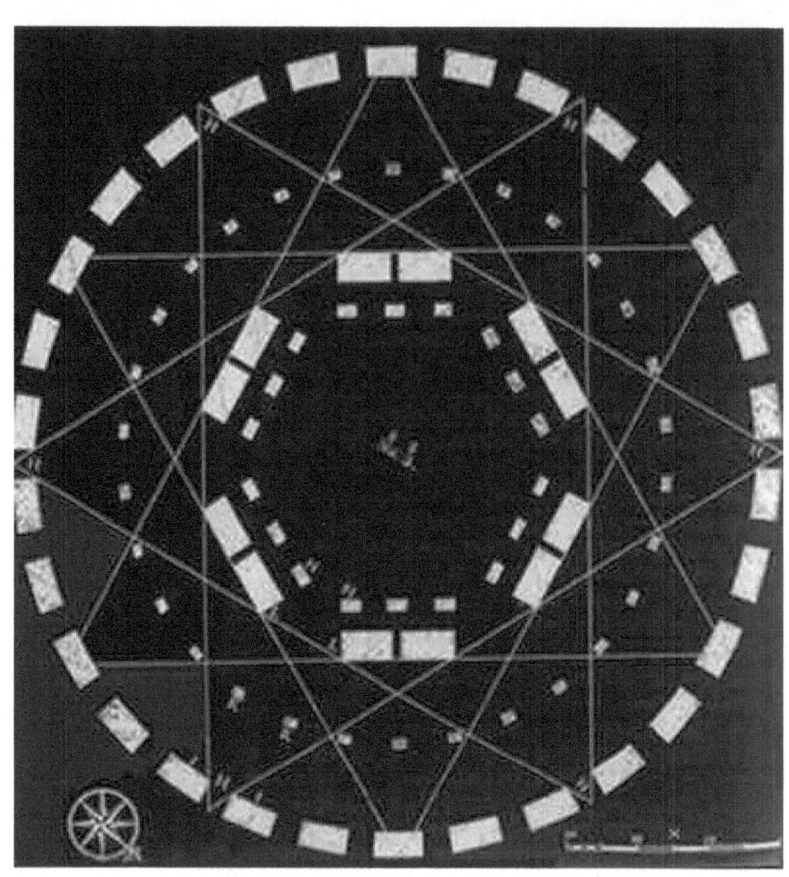

74

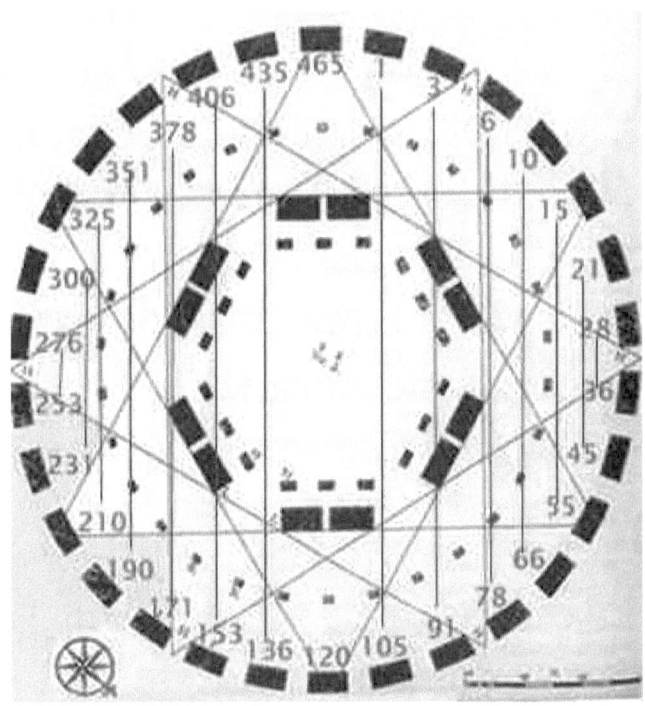

Do you see how a $STAR$ of
DAVID CONFIGURATION PATTERN could
fit into this circle of stone structures. Within the outer
circle is an inner HEXAGONAL configuration. Each
side of the HEXAGON has two stone(crystalline)
blocks(structures) side by side. Within that
HEXAGON is a slightly smaller HEXAGON with
each side of the HEXAGON accompanied by
THREE smaller stone(crystal) blocks(structures).
There is an obvious 2:3 ratio between the stones of the
inner HEXAGON with the structures of the
OUTER HEXAGON. 12 AND 18 ADD UP
TO 30 STONES(CRYSTALS)

or RESONATING STRUCTURES. This is is a 1 to 1 relationship with the 30 structures of the outer circle in this particular configuration pattern.

TETRAHEDRAL COFIGURATION of ENERGY FIELDS.

ELECTRON CONFIGURATION aROUND the ATOMIC NUCLUES.

Also if you look at the relationship of the triangle to the circle there is harmonic or numerical relationship between the triangle and the circle when triangular based numbers are placed within a circle in the same way that numbers are placed around a circle in a CLOCK. Starting with the number ONE the first 30 triangular numbers are 1,3,6,10,15,21,28,36,45,55,66,78,91,105,120,136,153,1 71,190,210,231,253,276,300,325,351, 378,406,435, AND 465. Look in the circle on the right they are also positioned on the circle. The circle is considered to be the most symmetrical two dimensional shape and the sphere the most symmetrical three dimensional shape. Anyway if you look at the diagram above on the RIGHT you will see how I linked the numbers up. The first line up of numbers to the right of the MIDLINE is 1 (at 12 degrees) and 105 (at 168 degrees). 105 plus 1 equals 106. Now look at the first line up of numbers to the left of the MIDLINE, there is435 (at 348 degrees) and 136 (at 192 degrees)435 plus 136 is 571. The difference between these two values 571 and 106 is 465which is the number at the TOP of the CIRCLE. Now look at the second line up of numbers to the right of the MIDLINE, 3 (at 24 degrees) and 91 (at 156 degrees)91 plus 3 equals 94. Now look at the second line up of numbers to the left of the MIDLINE, there is 406 (at 336 degrees) and 153 (at 204 degrees)406 plus 153 is 559. The difference between 559 and 94 is once again 465 the NUMBER at the TOP of the CIRCLE. Also if you take the number 435 (at 348 degress) and subtract it from 1 (at 12 degrees) you get434. Now go to other side and take 136 (at 192 degrees) and subtract it from 105 (at 168 degrees) and you get31. Now add 434 and 31 together you get 465--the 30th triangular number which is at the top of the circle. Clearly this pattern suggests that the triangular numbers must be a highly symmetrical grouping of numbers. Now what exactly is the significance or relationship of the TRIANGLE to the CIRCLE? Does a triangle or series of triangles inscribed within a circle serve as template or layout? Or a configuration pattern for energy and matter? Does it have something to do with the very basic or fundamental processes that govern the physics of matter and energy in this universe and perhaps other universes as well? Also if there is a fundamental harmonic relationship of the CIRCLE and TRIANGLE is there also a a fundamental harmonic relationship to the SPHERE and the TETRAHEDRON? Could this relationship have something to do with INTERDIMENSIONAL PHYSICS? A certain source had mentioned something about HYPERDIMENSIONAL ENERGY MODULATION.

Getting back to the CLOCK if you performed the calculations I have just done on the numbers ONE through TWELVE on an ordinary CLOCK you would get the same type of symmetries as with the triangular numbers. So clearly it seems that STONEHENGE was meant to serve as a CLOCK in some fashion . A CLOCK to monitor the CYCLES of the MOON. It seems evident that not all STONEHENGE is remaining any longer and there could have been quite a number of additional components in place at an earlier time which enabled a highly technical process to occur. Maybe the additional components, quite possibly the main component's could have been precisely machined stone or CRYSTALLINE structures that may have had some very unique or unusual internal geometrical properties that played a KEY roll in creating a specific configuration of energy fields.

HARMONIC ENERGY(SOUND) used in highly specific ways to RESONATE the STONE (CRYSTALLINE) STRUCTURES.

STONEHENGE STRUCTURES (FREQUENCIES)-EARTH---ENERGY GRID.

What is the nature of the geometrical configuration of
STONEHENGE?

ROTAING SOUND WAVES.
EXPERIMENT -CIRCULAR ARRAY of
SONIC GENERATORS.

VIBRATIONAL ENERGY
CYCLES of(PHASE) the SPATIAL
ENERGY FLUX..

THE USE OF PI OVER TWO RATIOS OF
FREQUENCIES.

The SARSEN stones and CENTRAL
TRILITHONS could be likened to giant
electromagnets which would set up interlocking
force(energy) fields around the center of this device. The
trilithon section remaining stationary(static field) in relation
to the SARSEN section which would spin around
it(RESONANT ROTATING FIELD). What

would result from the interaction of the
SARSEN(RESONANT ROTATING
FIELD) and the TRILITHON SECTION(static
field)? Would it be similar to the effect produced in
MAGNETIC RESONANCE in which
a RESONANT ROTATING FIELD(magnetic or
alternating) is imposed at right angles to a typically much
larger STATIC FIELD(electrostatic) to perturb the
orientation of nuclear moments(magnetic) in a piece of
matter. One source suggests that the interaction of the
ROTATING SARSEN FIELD with the electro-
STATIC TRILITHON FIELD would be to create
SPINNING MOTION.

One source suggested that the TRILITHON
SECTION because of the fact that the oval
arrangement is open at one end, would act as a magnetic jet
and give directional control to an aerial craft.

The BLUE STONE CIRCLE and the
AUBREY CIRCLE could possible represent wound
field coils much like the ones in an armature of an electric
motor(says one author). That is suggested by the
coupling of every second position in either circle. If these

field coils were then rotated in either direction in relation to the larger electromagnetic circles(toroid electromagnets) then spherical energy fields would be produced. If this were being used as a technology to launch a disc shaped space-craft then the spherical energy fields would encompass the whole of the vehicle. In this arrangement the Y and Z holes and the inner horse shoe electromagnetic device could represent some sort of controlling field arrangements that would provide stability and fine-tuning of the space-craft.

According to author BRUCE CATHIE all of the EM fields would be in perfect resonance with his unified equation and therefore with light itself.
By manipulating the energy fields through an infinite range of harmonics available, an interdimensional effect could be manifested which would allow travel in space-time.

Now there are of course other interesting possibilities for the use of this technology based on the STONEHENGE configuration pattern. As I have suggested before it is entirely possible that STONEHENGE or any similar device based on the same TEMPLATE could be used to generate energy fields based on a STAR OF DAVID geometrical configuration of crystalline structures. Maybe this STAR OF DAVID configuration pattern employed in a crystal grid structure and utilized in a highly specific way could be used to literally channel an enormous amount of higher-dimensional energy through the fabric of space into the center of the complex to form a pillar of light or some other massive concentration of energy. Or maybe if a huge amount of energy could be generated, possibly amplified and then focused into an extremely small area at the center of this circle then a bridge or gate could be created that would allow access to higher dimensional space possibly allowing for travel into other space-time continuums(universes) or also to act as a hyperspatial shortcut to far off distant locations in this universe.

SPINNING CRYSTAL LATTICE. INTERLOCKING GEOMETRICAL ENERGY FIELDS.

TETRAHEDRAL ENERGY PATTERNS(VORTEX?)

STONEHENGE as an ANCIENT
PARTICLE or ENERGY ACCELERATOR??
Whether this true or not the STONEHENGE
CONFIGURATION PATTERN which is
composed of a main outer ring and an inner ring could serve
as a template for the basic configuration of a particle
accelerator?

CONCLUSION

Could it be we aren't seeing something.

Could it be with all of our satellites, (which by the way, are synchronized to the Cartesian Co-ordinate system founded with Greenwich, England as the base point for the World Geodesic Survey System that is encoded in it) are all programmed incorrectly.

And all of the information that that the Global Positioning System Satellites are transmitting to your technological devices have to be re-calibrated?

However, which system do you use to recalibrate your technology?

Do you use one of the grid systems identified in this book?

Or something else?

Could there even be another grid system of a more ancient origin?

Could this system already be indicated by markers on the earth's surface?

Could Stonehenge be such a calibration device?

Are we blind to what is already here?

Stonehenge

Stonehenge is a prehistoric monument located in the English county of Wiltshire, about 3.2 kilometres (2.0 mi) west of Amesbury and 13 kilometres (8.1 mi) north of Salisbury. One of the most famous prehistoric sites in the world, Stonehenge is composed of earthworks surrounding a circular setting of large standing stones. Archaeologists had believed that the iconic stone monument was erected around 2500 BC. However one recent theory has suggested that the first stones were not erected until 2400-2200 BC, while another suggests that bluestones may have been erected at the site as early as 3000 BC.

The surrounding circular earth bank and ditch, which constitute the earliest phase of the monument, have been dated to about 3100 BC. The site and its surroundings were added to the UNESCO's list of World Heritage Sites in 1986 in a co-listing with Avebury henge monument, and it is also a legally protected Scheduled Ancient Monument. Stonehenge itself is owned by the Crown and managed by English Heritage while the surrounding land is owned by the National Trust.

New archaeological evidence found by the Stonehenge Riverside Project indicates that Stonehenge served as a burial ground from its earliest beginnings. The dating of cremated remains found that burials took place as early as 3000 B.C, when the first ditches were being built around the monument. Burials continued at Stonehenge for at least another 500 years when the giant stones which mark the landmark were put up.

History

Stonehenge itself evolved in several construction phases spanning at least some 1500 years. However there is evidence of large scale construction both before and afterwards on and around the monument that perhaps extends the landscape's time frame to 6500 years.

Dating and understanding the various phases of activity at Stonehenge is not a simple task; it is complicated by poorly kept early excavation records, surprisingly few accurate scientific dates and the disturbance of the natural chalk by periglacial effects and animal burrowing. The modern phasing most generally agreed by archaeologists is detailed below. Features mentioned in the text are numbered and shown on the plan, right, which illustrates the site as of 2004. The plan omits the trilithon lintels for clarity. Holes that no longer, or never, contained stones are shown as open circles and stones visible today are shown colored. It is widely assumed that Stonehenge once stood as a magnificent "complete" monument; we should be aware that this cannot be proved, since around half of the stones that should be present are in fact missing, and since many of the assumed stone sockets have never actually been recorded through excavation.

Before the monument (8000 BC forward)

Some archaeologists have found four (or possibly five, although one may have been a natural tree throw) large Mesolithic postholes which date to around 8000 BC nearby, beneath the modern tourist car-park. These held pine posts around 0.75 metres (2.5 ft) in diameter which were erected and left to rot in situ. Three of the posts (and possibly four) were in an east-west alignment and may have had ritual significance; no parallels are known from Britain at the time but similar sites have been found in Scandinavia. At this time, Salisbury Plain was still wooded but four thousand years later, during the earlier Neolithic, a causewayed enclosure at Robin Hood's Ball and long barrow tombs were built in the surrounding landscape. In approximately 3500 BC a large cursus monument was built 700 metres (2,300 ft) north of the site as the first farmers began to clear the forest and exploit the area.

Stonehenge 1 (ca. 3100 BC)

The first monument consisted of a circular bank and ditch enclosure made of Late Cretaceous (Santonian Age) Seaford Chalk, (7 and 8) measuring around 110 metres (360 ft) in diameter with a large entrance to the north east and a smaller one to the south (14). It stood in open grassland on a slightly sloping but not especially remarkable spot. The builders placed the bones of deer and oxen in the bottom of the ditch as well as some worked flint tools. The bones were considerably older than the antler picks used to dig the ditch and the people who buried them had looked after them for some time prior to burial. The ditch itself was continuous but had been dug in sections, like the ditches of the earlier causewayed enclosures in the area.

The chalk dug from the ditch was piled up to form the bank. This first stage is dated to around 3100 BC after which the ditch began to silt up naturally and was not cleared out by the builders. Within the outer edge of the enclosed area was dug a circle of 56 pits, each around 1 metre (3.3 ft) in diameter (13), known as the Aubrey holes after John Aubrey, the seventeenth century antiquarian who was thought to have first identified them. The pits may have contained standing timbers, creating a timber circle although there is no excavated evidence of them. A recent excavation has suggested that the Aubrey Holes may have originally been used to erect a bluestone circle.[6] If this were the case it would advance the earliest known stone structure at the monument by some 500 years. A small outer bank beyond the ditch could also date to this period.

Stonehenge 2 (ca. 3000 BC)

Evidence of the second phase is no longer visible. It appears from the number of postholes dating to this period that some form of timber structure was built within the enclosure during the early 3rd millennium BC. Further standing timbers were placed at the northeast entrance and a parallel alignment of posts ran inwards from the southern entrance. The postholes are smaller than the Aubrey Holes, being only around 0.4 metres (16 in) in diameter and are much less regularly spaced. The bank was purposely reduced in height and the ditch continued to silt up. At least twenty-five of the Aubrey Holes are known to have contained later, intrusive, cremation burials dating to the two centuries after the monument's inception.

It seems that whatever the holes' initial function, it changed to become a funerary one during Phase 2. Thirty further cremations were placed in the enclosure's ditch and at other points within the monument, mostly in the eastern half. Stonehenge is therefore interpreted as functioning as an enclosed cremation cemetery at this time, the earliest known cremation cemetery in the British Isles. Fragments of unburnt human bone have also been found in the ditch fill. Late Neolithic grooved ware pottery has been found in connection with the features from this phase providing dating evidence.

Stonehenge 3 I (ca. 2600 BC)

Archaeological excavation has indicated that around 2600 BC, timber was abandoned in favour of stone, and two concentric arrays of holes (the Q and R Holes) were dug in the centre of the site. These stone sockets are only partly known (hence on present evidence are sometimes described as forming Œcrescents¹), however they could be the remains of a double ring. Again, there is little firm dating evidence for this phase. The holes held up to 80 standing stones (shown blue on the plan) only 43 of which can be traced today.

About 2,000 BC, the first stone circle (which is now the inner circle), comprised of small bluestones, was set up, but abandoned before completion. The stones used in that first circle are believed to be from the Prescelly Mountains, located roughly 240 miles away, at the southwestern tip of Wales. The bluestones weigh up to 4 tons each and about 80 stones were used, in all. Given the distance they had to travel, this presented quite a transportation problem.

Modern theories speculate that the stones were dragged by roller and sledge from the inland mountains to the headwaters of Milford Haven. There they were loaded onto rafts, barges or boats and sailed along the south coast of Wales, then up the Rivers Avon and Frome to a point near present-day Frome in Somerset. From this point, so the theory goes, the stones were hauled overland, again, to a place near Warminster in Wiltshire, approximately 6 miles away. From there, it's back into the pool for a slow float down the River Wylye to Salisbury, then up the Salisbury Avon to West Amesbury, leaving only a short 2 mile drag from West Amesbury to the Stonehenge site.

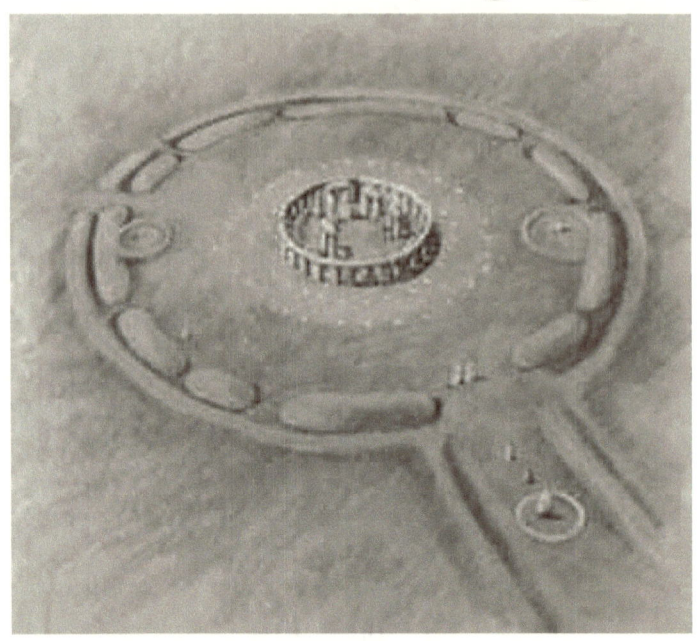

The bluestones (some of which are made of dolerite, an igneous rock), were thought for much of the 20th century to have been transported by humans from the Preseli Hills, 250 kilometres (160 mi) away in modern day Pembrokeshire in Wales. A newer theory is that they were brought from glacial deposits much nearer the site, which had been carried down from the northern side of the Preselis to southern England by the Irish Sea Glacier.

Other standing stones may well have been small sarsens, used later as lintels. The stones, which weighed about four tons, consisted mostly of spotted Ordovician dolerite but included examples of rhyolite, tuff and volcanic and calcareous ash; in total around 20 different rock types are represented. Each monolith measures around 2 metres (6.6 ft) in height, between 1 m and 1.5 m (3.3-4.9 ft) wide and around 0.8 metres (2.6 ft) thick. What was to become known as the Altar Stone (1), is is almost certainly derived from either Carmarthenshire or the Brecon Beacons and may have stood as a single large monolith.

The north eastern entrance was also widened at this time with the result that it precisely matched the direction of the midsummer sunrise and midwinter sunset of the period. This phase of the monument was abandoned unfinished however, the small standing stones were apparently removed and the Q and R holes purposefully backfilled. Even so, the monument appears to have eclipsed the site at Avebury in importance towards the end of this phase. The Heelstone (5), a Tertiary sandstone, may also have been erected outside the north eastern entrance during this period although it cannot be securely dated and may have been installed at any time in phase 3.

At first, a second stone, now no longer visible, joined it. Two, or possibly three, large portal stones were set up just inside the north eastern entrance of which only one, the fallen Slaughter Stone (4), 4.9 metres (16 ft) long, now remains. Other features loosely dated to phase 3 include the four Station Stones (6), two of which stood atop mounds (2 and 3). The mounds are known as 'barrows' although they do not contain burials. The Avenue, (10), a parallel pair of ditches and banks leading 3 kilometres (1.9 mi) to the River Avon was also added. Two ditches similar to Heelstone Ditch circling the Heelstone, which was by then reduced to a single monolith, were later dug around the Station Stones.

Stonehenge 3 II (2600 BC to 2400 BC)

The next major phase of activity saw 30 enormous Oligocene-Miocene sarsen stones (shown grey on the plan) brought to the site. They may have come from a quarry around 40 kilometres (25 mi) north of Stonehenge, on the Marlborough Downs, or they may have been collected from a "litter" of sarsens on the chalk downs, closer to hand. The stones were dressed and fashioned with mortise and tenon joints before 30 were erected as a 33 metres (110 ft) diameter circle of standing stones, with a ring of 30 lintel stones resting on top.

The lintels were fitted to one another using another woodworking method, the tongue and groove joint. Each standing stone was around 4.1 metres (13 ft) high, 2.1 metres (6.9 ft) wide and weighed around 25 tons. Each had clearly been worked with the final effect in mind; the orthostats widen slightly towards the top in order that their perspective remains constant as they rise up from the ground while the lintel stones curve slightly to continue the circular appearance of the earlier monument.

The sides of the stones that face inwards are smoother and more finely worked than the sides that face outwards. The average thickness of these stones is 1.1 metres (3.6 ft) and the average distance between them is 1 metre (3.3 ft). A total of 74 stones would have been needed to complete the circle and unless some of the sarsens were removed from the site, it would seem that the ring was left incomplete. Of the lintel stones, they are each around 3.2 metres (10 ft), 1 metre (3.3 ft) wide and 0.8 metres (2.6 ft) thick. The tops of the lintels are 4.9 metres (16 ft) above the ground.

Within this circle stood five trilithons of dressed sarsen stone arranged in a horseshoe shape 13.7 metres (45 ft) across with its open end facing north east. These huge stones, ten uprights and five lintels, weigh up to 50 tons each and were again linked using complex jointing. They are arranged symmetrically; the smallest pair of trilithons were around 6 metres (20 ft) tall, the next pair a little higher and the largest, single trilithon in the south west corner would have been 7.3 metres (24 ft) tall. Only one upright from the Great Trilithon still stands; 6.7 metres (22 ft) is visible and a further 2.4 metres (7.9 ft) is below ground.

The images of a 'dagger' and 14 'axe-heads' have been recorded carved on one of the sarsens, known as stone 53. Further axe-head carvings have been seen on the outer faces of stones known as numbers 3, 4, and 5. They are difficult to date but are morphologically similar to later Bronze Age weapons; recent laser scanning work on the carvings supports this interpretation. The pair of trilithons in north east are smallest, measuring around 6 metres (20 ft) in height and the largest is the trilithon in the south west of the horseshoe is almost 7.5 metres (25 ft) tall. This ambitious phase is radiocarbon dated to between 2600 and 2400 BC[8]. This is slightly before two sets of burials discovered 3 miles (4.8 km) to the west in Amesbury (the Amesbury Archer found in 2002, and the Boscombe Bowmen discovered in 2003) as well as the Stonehenge Archer whose body was discovered in the outer ditch of the monument in 1978.

At a similar time a large timber circle and another avenue were constructed overlooking the River Avon 2 miles away at Durrington Walls. Opposing the solar alignments at Stonehenge, the circle was orientated towards the rising sun on the midwinter solstice, whilst the Avenue led from the river to the circle on an alignment to the setting sun on the summer solstice. Evidence of huge fires on the banks of the Avon between the two avenues also suggests that both circles were linked, and perhaps formed a procession route used on the longest and shortest days of the year. Parker Pearson speculates that the wooden circle at Durrington Walls was the centre of a 'land of the living', whilst the stone circle represented a 'land of the dead'. The Avon would have served as a journey between the two.

Stonehenge 3 III

Later in the Bronze Age, the bluestones appear to have been re-erected for the first time, although the exact details of this period are still unclear. They were placed within the outer sarsen circle and at this time may have been trimmed in some way. A few have timber working-style cuts in them like the sarsens themselves, suggesting they may have been linked with lintels and part of a larger structure during this phase.

Stonehenge 3 IV (2280 BC to 1930 BC)

This phase saw further rearrangement of the bluestones as they were placed in a circle between the two settings of sarsens and in an oval in the very centre. Some archaeologists argue that some of the bluestones in this period were part of a second group brought from Wales. All the stones were well-spaced uprights without any of the linking lintels inferred in Stonehenge 3 III. The Altar Stone may have been moved within the oval and stood vertically. Although this would seem the most impressive phase of work, Stonehenge 3 IV was rather shabbily built compared to its immediate predecessors, as the newly re-installed bluestones were not at all well founded and began to fall over. However, only minor changes were made after this phase. Stonehenge 3 IV dates from 2280 to 1930 BC.

Stonehenge 3 V (2280 BC to 1930 BC)

Soon afterwards, the north eastern section of the Phase 3 IV Bluestone circle was removed, creating a horseshoe-shaped setting termed the Bluestone Horseshoe. This mirrored the shape of the central sarsen Trilithons and dates from 2270 to 1930 BC. This phase is contemporary with the famous Seahenge site in Norfolk.

After the monument (1600 BC on)

The last known construction at Stonehenge was about 1600 BC (see 'Y and Z Holes' below), and the last known usage of it was likely during the Iron Age. Roman coins and medieval artifacts have all been found in or around the monument but it is unknown if the monument was in continuous use throughout prehistory and beyond - or exactly how it would have been used. Notable is the late 7th-6th century BC large arcing Scroll Trench which deepens E-NE towards Heelstone, and the construction of the massive Iron Age hillfort Vespasian's Camp built alongside the Avenue near the Avon. The burial of a decapitated 7th century Saxon man was excavated from Stonehenge. The site was known by scholars during the Middle Ages and since then it has been studied and adopted by numerous different groups.

16th to 20th centuries

Stonehenge has changed ownership on several occasions since King Henry VIII acquired Amesbury Abbey and its surrounding lands. In 1540 he gave the estate to the Earl of Hertford, and it subsequently passed to Lord Carleton and then the Marquis of Queensbury. The Antrobus family of Cheshire bought the estate in 1824, but sold it in 1915 after the last heir was killed serving in France during the First World War. The auction was held by Knight Frank & Rutley estate agents in Salisbury on the 21 September, and included "Lot 15. Stonehenge with about 30 acres, 2 rods, 37 perches of adjoining downland." Cecil Chubb bought Stonehenge for 6,600 pds. and then gave it to the nation three years later. Although it has been speculated that he purchased it at the suggestion of - or even as a present for - his wife, he in fact bought it on a whim as he believed a local man should be the new owner.

1920s onwards

In the late 1920s a nation-wide appeal was launched to save Stonehenge from the encroachment of modern buildings that had begun to appear around it. During World War I an aerodrome had been built on the down just west of the circle, and in the dry valley at Stonehenge Bottom a main road junction had appeared, with several cottages and a cafe. In 1928 the land around the stones was purchased with the appeal donations, and given to the National Trust in order to preserve it. The buildings were removed (although the roads were not), and the land returned to agriculture. More recently the land has been part of a grassland reversion scheme, returning the surrounding fields to native chalk grassland.

In 2002 a public poll voted Stonehenge as one of the seven wonders of Britain, alongside Big Ben, the Eden Project, Hadrian's Wall, the London Eye, Windsor Castle, and York Minster.

As motorised traffic increased the setting of the monument began to be affected by the proximity of the two roads on either side of it - the A344 to Shrewton on the north side, and the A303 to Winterbourne Stoke to the south. Plans to upgrade the A303 and remove it from the view of the stones have been considered since it became a World Heritage Site, but the controversy surrounding expensive re-routings of a road have led to the scheme being cancelled on multiple occasions. On 06 December 2007 it was announced that the most recent plans had been cancelled.

Neopaganism

Stonehenge is a place of pilgrimage for neo-druids, and for certain others following pagan or neo-pagan beliefs. The midsummer sunrise began attracting modern visitors in the 1870s, with the first record of recreated Druidic practices dating to 1905 when the Ancient Order of Druids enacted a ceremony. Despite efforts by archaeologists and historians to stress the differences between the Iron Age Druidic religion and the much older monument, Stonehenge has become increasingly, almost inextricably, associated with British Druidism, Neopaganism and New Age philosophy. Between 1972 and 1984, Stonehenge was the site of a free festival. After the Battle of the Beanfield in 1985 this use of the site was stopped for several years, and currently ritual use of Stonehenge is carefully controlled.

Access

When Stonehenge became open to the public it was possible to walk amongst and even climb on the stones. However this ended in 1977 when the stones were roped off as a result of serious erosion. Visitors are no longer permitted to touch the stones, but merely walk around the monument from a short distance. English Heritage does however permit access during the summer and winter solstice, and the spring and autumn equinox. Additionally, visitors can make special bookings to access the stones throughout the year.

Archaeological Research and Restoration

Throughout recorded history Stonehenge and its surrounding monuments have attracted attention from antiquarians and archaeologists. John Aubrey was one of the first to examine the site with a scientific eye in 1666, and recorded in his plan of the monument the pits that now bear his name. William Stukeley continued Aubrey's work in the early 18th century, but took an interest in the surrounding monuments as well, identifying (somewhat incorrectly) the Cursus and the Avenue. He also began the excavation of many of the barrows in the area, and it was his interpretation of the landscape that associated it with the Druids. Stukeley was in fact so fascinated with Druids that he originally named Disc Barrows as Druids Barrows. The most accurate early plan of Stonehenge was that made by Bath architect John Wood in 1740. His original annotated survey has recently been computer redrawn and published. Importantly Wood's plan was made before the collapse of the southwest Trilithon (which fell in 1797; restored 1958).

William Cunnington was the next to tackle the area in the early 19th century, excavating some 24 barrows before digging in and around the stones, discovering charred wood, animal bones, pottery and urns. He also identified the hole in which the Slaughter Stone once stood. At the same time Richard Colt Hoare began his activities, excavating some 379 barrows on Salisbury Plain before working with Cunnington and William Coxe on some 200 in the area around the Stones. To alert future diggers to their work they were careful to leave initialed metal tokens in each barrow they opened.

In 1877 Charles Darwin dabbled in archaeology at the stones, experimenting with the rate at which remains sink into the earth for his book The Formation of Vegetable Mould Through the Action of Worms.

William Gowland oversaw the first major restoration of the monument in 1901 - the straightening and concrete setting of sarsen stone number 56 which was in danger of falling. Unfortunately in straightening it he also moved it about half a metre from its original position. He also took the opportunity to further excavate the monument at the same time in what was the most scientific dig to date, revealing more about the erection of the stones than the previous 100 years of work.

During the 1920 restoration William Hawley, who had excavated nearby Old Sarum, excavated the base of six stones being restored as well as the outer ditch. He also located a bottle of port in the slaughter stone socket left by Cunnington, helped to rediscover Aubrey's pits inside the bank and located the Y and Z Holes (concentric circular holes outside the Sarsen Circle). Richard Atkinson, Stuart Piggott and John F. S. Stone re-excavated much of Hawley's work in the 40s and 50s, and discovered the carved axes and daggers on the Sarsen Stones. Atkinson's work was instrumental in the understanding of the three major phases of the monument's construction.

In 1958 the stones were restored again, using concrete settings to re-erect three of the standing sarsens. The very last restoration was carried out in 1963 when stone 23 of the Sarsen Circle fell over and was once more re-erected, and the opportunity taken to concrete three more stones. Later archaeologists, including Christopher Chippindale of the Museum of Archaeology and Anthropology, University of Cambridge and Brian Edwards of the University of the West of England campaigned to give the public more knowledge of the various restorations and in 2004 English Heritage included pictures of the works in progress in its new book Stonehenge: A History in Photographs.

In 1966 and 1967, in advance of a new car park being built at the site, the area of land immediately northwest of the stones was excavated by Faith and Lance Vatcher. They discovered the Mesolithic postholes dating from between 7 and 8,000 BC, as well as a 10m length of a palisade ditch - a V cut ditch into which timber posts had been inserted that remained there until they rotted away. Subsequent aerial archaeology suggests that this ditch runs from the west to the north of Stonehenge, near the avenue.

Excavations were once again carried out in 1978 by Atkinson and John Evans during which they discovered the remains of the Stonehenge Archer from the outer ditch, and in 1979 rescue archaeology was needed alongside the Heel Stone after a cable-laying ditch was mistakenly dug on the roadside, revealing a new stone hole next to the Heel Stone.

In the early 1980's Julian Richards led the Stonehenge Environs Project, a detailed study of the surrounding landscape. The project was able to successfully date such features as the Lesser Cursus, Coneybury henge and several other smaller features.

More recent excavations include Mike Parker Pearson's Stonehenge Riverside Project - an series of digs held between 2003 and 2008. This project mainly investigated other monuments in the landscape and their relationship with the stones - notably Durrington Walls where another Avenue leading to the river Avon was discovered.

In April 2008 Professor Tim Darvill of the University of Bournemouth and Professor Geoff Wainwright of the Society of Antiquaries began another dig inside the Stone circle to retrieve dateable fragments of the original bluestone pillars. They were able to date the erection of some bluestones to 2300BC, although this may not reflect the earliest erection of stones at Stonehenge. They also discovered organic material from 7000 B.C., which, along with the Mesolithic postholes, adds support for the site having been in use at least 4000 years before Stonehenge was started. In August and September 2008, as part of the Riverside Project Julian Richards and Mike Pitts excavated Aubrey Hole 7, removing the cremated remains from several Aubrey Holes that had been excavated by Hawley in the 1920s, and re-interred in 1935

These are just some of the possibilities that come to mind.

Maybe when you read this text you might open up to something,..........

Something more.

Happy Musings.

t. gyurme

Tenzins other books include:

'Zero to Mastery in 7 Volumes'

The Mastery Series consisting of Seven Books

'Our Christening'

'S-Alchemy'

'Buddhism and God'

'Thoth, God of Moon, Magic and Writing'

'The Flower of Life and The History of Meta-Physics'

'Contemplative Meta-Physical Healing'

'I Am, In a Post-Modern World'

These books and more information about trainings from TLC Training are available at www.thezenden.eu, and Tenzin is available for individual coaching at tzingyurme@gmail.com.

www.ingramcontent.com/pod-product-compliance
Lightning Source LLC
Chambersburg PA
CBHW030842180526
45163CB00004B/1427